Modeling Dynamic Systems

Series Editors

Matthias Ruth
Bruce Hannon

For further volumes:
http://www.springer.com/series/3427

Matthias Ruth • Bruce Hannon

Modeling Dynamic Economic Systems

Second Edition

Foreword by Jay W. Forrester

 Springer

Matthias Ruth
University of Maryland
College Park, MD 20742, USA

Bruce Hannon
University of Illinois
Urbana, IL 61801, USA

A save-disabled version of STELLA® and the computer models of this book are available at
www.iseesystems.com/modelingeconomicsystems.

ISBN 978-1-4614-2208-2   e-ISBN 978-1-4614-2209-9
DOI 10.1007/978-1-4614-2209-9
Springer New York Dordrecht Heidelberg London

Library of Congress Control Number: 2012930135

Printed on acid-free paper

Springer is part of Springer Science+Business Media (www.springer.com)

Foreword to the First Edition

The frontiers of physical science and technology commanded attention during the last two centuries. In the next century lies the frontier of better understanding behavior of social and economic systems. Ruth and Hannon join many other pioneers exploring this new frontier.

By building on the profession of system dynamics, Ruth and Hannon contribute to movement away from purely mental models, which necessarily lack adequate precision, toward more insightful and disciplined computer simulation models. The STELLA software used in this book is one of several computer applications created to implement the concepts of system dynamics, a discipline that has been developed over the last 40 years and now extends into many fields of activity. System dynamics is beginning, already even in kindergarten through 12th grade education, to provide a rigorous foundation for dealing with dynamic change in mathematics, physics, social studies, environment, history, and even literature. Education at every level will be changing from teaching isolated facts to allowing students to explore those systems within which facts, policies, and individual relationships are imbedded to develop their ability to think in terms of dynamic systems.

This book includes elementary instruction in system dynamics modeling and in the STELLA software. It covers a wide range of material from simple building blocks of systems to models and mathematics of considerable complexity. The material opens many avenues for further exploration, refinement, and simplification. Examples demonstrate the process of moving from mental models to computer simulation models. Mental models contain a wealth of useful information about policies and relationships in a system, but mental intuition cannot reliably estimate the future dynamic consequences when those elementary relationships interact with one another. Converting mental models to computer simulation models forces a clarity of thought that is missing in ordinary writing and discussion. And, most important, computer models reliably reveal the behavior implied by the structure of a system that has been described. The authors are to be commended for pushing toward a much more meaningful way to deal with the world's major difficulties in social and economic systems.

Jay W. Forrester

Preface

The problems of understanding complex system behavior and the challenge of developing easy-to-use models are apparent in the fields of environmental management and economic development. We are faced with the problem of reconciling economic activities with the goal of preserving or increasing environmental quality and quality of life. In economic and environmental problems, many parameters need to be assessed. This requires tools that enhance the collection and organization of data, interdisciplinary model development, transparency of models, and visualization of the results. Neither purely mathematical nor purely experimental approaches will suffice to help us better understand the world we live in and shape so intensively.

Until recently, we needed significant preparation in mathematics and computer programming to develop such models. Because of this hurdle, many have failed to give serious consideration to preparing and manipulating computer models of the events and developments in the world around them. This book, and the methods on which it is built, will empower us to model and analyze the dynamics characteristic of economic processes and human–environment interactions.

Without computer models, we are often left to choose between two strategies. First, we may resort to theoretical models that describe the world around us. Mathematics offers powerful tools for such descriptions, adhering to logic and providing a common language by sharing similar symbols and tools for analysis. Mathematical models are appealing in social and natural science, where cause and effect relationships are confusing. These models, however, run the risk of becoming detached from reality, sacrificing realism for analytical tractability. As a result, these models are accessible only to the trained scientist, leaving others to "believe or not believe" the model results.

Second, we may manipulate real systems to understand cause and effect. One could modify the system experimentally (e.g., introduce a pesticide, some CO_2, etc.) and observe the effects. If no significant effects are noted, one is free to assume the action has no effect and increase the level of the system change. This is an exceedingly common approach. It is an elaboration of the way an auto mechanic repairs an engine, by trial and error. But social and ecological systems are not auto engines. Errors in tampering with these systems can have substantial costs, in both the short term and the long term, and can bring with them a series of other unintended and unanticipated consequences that may lead to more tampering, further errors, and so on. Despite growing evidence, the trial and error approach remains the meter of the day. We trust that, just like the auto mechanic, we will be clever enough to clear up the problems created by the introduced change. We let our tendency toward optimism mask the new problems.

However, the level of intervention in social and ecological systems has become so great that the adverse effects cannot be ignored. As our optimism about repair begins to crumble, we take on the attitude of patience toward the inevitable— unassignable cancer risk, global warming, fossil fuel depletion—the list is long. We are pessimistic about our ability to identify and influence cause and effect relationships. We need to understand the interactions of the components of dynamic systems in order to guide our actions. We need to add synthetic thinking to the reductionist approach. Otherwise, we will continue to be overwhelmed by details, failing to see the forest for the trees.

There is something useful that we can do to turn from this path. We can experiment using computer models. Models give us predictions of the short- and long-term outcomes of proposed actions. To do this we can effectively combine mathematical models with experimentation. By building on the strengths of each we will gain insight that exceeds the knowledge derived from choosing one method over the other. Experimenting with computer models will open a new world in our understanding of dynamic systems. The consequences of discovering adverse effects in a computer model are no more than ruffled pride.

Computer modeling has been with us for nearly 50 years. Why then are we so enthusiastic about its use now? The answer comes from innovations in software and powerful, affordable hardware available to every individual. Almost anyone can now begin to simulate real-world phenomena on his or her own, in terms that are easily explainable to others. Computer models are no longer confined to the computer laboratory. They have moved into every classroom, and we believe they can and should move into the personal repertoire of every educated citizen.

The ecologist Garrett Hardin and the physicist Heinz Pagels have noted that an understanding of system function, as a specific skill, needs to be and can become an integral part of general education. It requires the recognition (easily demonstrable with exceedingly simple computer models) that the human mind is not capable of handling very complex dynamic models by itself. Just as we need help in seeing bacteria and distant stars, we need help modeling dynamic systems. We do solve the crucial dynamic modeling problem of ducking stones thrown at us or safely crossing busy streets. We learned to solve these problems by being shown the logical outcome of mistakes or through survivable accidents of judgment. We experiment with the real world as children and get hit by hurled stones, or we let adults play out their mental model of the consequences for us and we believe them. These actions are the result of experimental and predictive models, and they begin to occur at an early age. In the complex social, economic, and ecological world, however, we cannot rely on the completely mental model for individual or especially for group action, and often we cannot afford to experiment with the system in which we live. We must learn to simulate, experiment, and predict with complex models.

In this book, we have selected the modeling software STELLA® with its iconographic programming style. Programs such as STELLA are changing the way in which we think. They enable each of us to focus and clarify the mental model we have of a particular phenomenon, to augment it, to elaborate it, and then to do something we cannot otherwise do: to run it, to let it yield the inevitable dynamic consequences hidden in our assumptions and the structure of the model. STELLA is not the ultimate tool in this process of mind extension. However, its relative ease of use makes the path to freer and more powerful intellectual inquiry accessible to every student.

These are the arguments for our book on *Dynamic Modeling of Natural Resource Use*. This volume was spurred by our first book on *Dynamic Modeling,* (Hannon and Ruth 1994) and others in the series on *Modeling Dynamic Systems*. The need to enhance our knowledge of human–environment interactions and the recognition that traditional teaching of economy–environment interactions frequently lacks tools that enable students to investigate, through an experimental approach, their own understanding of these interactions and to compare their findings against reality. We consider such modeling as the most important task before us. To help students learn to extend the reach of their minds in this unfamiliar yet very powerful way is the most important thing we can do.

College Park, MD, USA   Matthias Ruth
Urbana, IL, USA   Bruce Hannon

Reference

Hannon B, Ruth M (1994) Dynamic modeling, 1st edn (2nd edn, 2001). Springer, New York

Notation

Term	Typical units	Symbol
Resource stock	Tons	X
Input quantity	Tons/year	Y
Production, output	Tons/year	Q
Total cost	Dollars/year	C
Average cost	Dollars/ton	AC
Marginal cost	Dollars/ton	MC
Input price	Dollars/ton	W
Output price	Dollars/ton	P
Revenue	Dollars/year	R
Marginal revenue	Dollars/ton	MR
Interest rate	Dollars/dollar/year	I
Current value of profit	Dollars/year	CVP
Present value of profit	Dollars/year	PVP
Annual payment	Dollars/year	AP
Cumulative current value of profit	Dollars	CCVP
Cumulative present value of profit	Dollars	CPVP
Cumulative production	Tons	Z
Discovery rate	Tons/year	D
Discovery cost	Dollars/ton	DC
Marginal discovery cost	Dollars/year	MDC

Contents

Part I
Introduction

Chapter 1
Introduction

*People confident in their own objectivity are likely to be
arrogant and self deceived—less likely, in fact, to see things
as clearly as those who recognize and acknowledge their own
biases. Those who have faith in the objectivity of others may
be complacent or dangerously naive. They fail to see the
many obstacles, innate and acquired, innocent and insidious,
inevitable and unavoidable, along the road to truth.*

Lichtenberg, J. 1991. Objectivity and Its Enemies,
The Responsive Community, Vol. 2, pp. 59–69

1.1 Modeling Dynamic Systems

1.1.1 Model Building

This book was borne out of our understanding that arrogance, self-deception, and
blind reliance in science on the truths of others can be eradicated only if we extend
our intellectual capacities to embrace and effectively communicate the richness of
our real-world experiences. Our book attempts to show a path to freer, system-based
thinking and to encourage you to challenge yourself and the wisdom of others.

Models are central to our understanding of the world because they enable us to
represent and manipulate real phenomena, then explore the results. We create
models of cause and effect every day, building on past experience to evaluate
present options. For instance, you determine the "best" route to a store based on
your experience of its surroundings. Scientists and theoreticians also build on their
personal experience of the real world in their research, but they do so in a more

---

A save-disabled version of STELLA® and the computer models of this book are available at
www.iseesystems.com/modelingeconomicsystems.

M. Ruth and B. Hannon, *Modeling Dynamic Economic Systems*,
Modeling Dynamic Systems, DOI 10.1007/978-1-4614-2209-9_1,
© Springer Science+Business Media, LLC 2012

organized and formal way. Among them are economists who model how humans meet their needs with given endowments of resources and technologies.

A large number of models have been developed by the economics profession over the last 200 years. Each is concerned with some part of the economy and makes different assumptions on the processes that influence that part. Recently, increasing attention has been given to the relationships between the economy and the environment. Ultimately, all the materials and energy used to meet economic needs must be extracted from the environment. Extraction, processing, distribution, and use of materials and energy, in turn, lead to the generation of waste. Both resource depletion and waste generation are likely to affect human welfare and therefore need to be understood to make proper economic decisions.

This book deals with the dynamic processes in economic systems and concentrates on the extraction and use of natural resources. We develop models to better understand the impact of alternative decisions on economic performance and environmental quality. Models are essential tools in generating new knowledge. They help us simplify complex phenomena by eliminating everything we believe is extraneous to what we want to study. An equation is a model stripped of everything but the relationships between the variables and some initial conditions of the system under investigation. Computers allow us to expand the scope of our models, to include more and more diverse variables, and to ask more "what-if" questions. Heinz Pagel compared the insight given by computer models to the extended sight given by microscopes and telescopes (Pagels 1988). With the right models we can explore the replication of viral particles or the economy of a nation, and we can explore dynamic as well as static phenomena. Through such knowledge, we can explain the world around us and, possibly, anticipate future happenings.

Real-world phenomena can be difficult to study. We are easily overwhelmed with a flood of details, complex interrelationships, and a seeming lack of organization, as well as the dynamic nature of reality in process. To deal with such complex events, we formulate questions about the process we want to study and try to answer these questions. We look for patterns within the details, without losing sight of the "big picture"—we search for the underlying key, the particular set of details or structure that leads to some outcome we have seen. Models, as abstractions of reality, force us to consider the results of our structural and dynamic assumptions.

Model building can be an involved process, but certain procedures are generally followed. Suppose a real event sparks your curiosity about what caused it or what it could lead to. You translate this curiosity into questions or a theory about the event and the processes surrounding it. You identify key elements, and your assumptions about them do form an abstract version of the event. For example, in determining the "best" route to store, you would take into account its geographic location, the time it opens, the traffic, and other obstacles that may impede your speed to get there, and your previous location, the time you have to make the trip, and possible routes for your walk or drive. You consider the weather only if it would affect your speed between the two locations. The places you would pass along the way will not enter into your determination, unless you want to make a conscious choice about combining a visit to them with your visit to the store. Probably, you abstract away from them. The parameters left for you to determine are distance, the time you have to traverse it, possible paths, and factors that might affect your speed.

After you narrow down the details to those that describe the problem, you must specify the relationships among them. The relevant details (the variables) and their relationships establish your model. You should be able to tell if you can get to the store at a good time.

"Running" the model—in this case, relating distance, speed, and possible obstacles—gives you the experience and data on which to draw conclusions and, perhaps, predict future experiences. Such conclusions and expectations, then, can be compared to other events you have experienced. In the light of real-world experiences, you might decide to reject the model, accept it, or most likely, revise it. Model building and revision is dynamic—build, run, compare, and change—and each cycle improves your understanding of reality. Sometimes, the modeling process forces you to recognize new and important parameters; sometimes, modeling will make you reduce the importance of your favorite parameter.

According to Casti (1989), good models are the simplest ones that explain the data and yet do not explain it all, leaving some room for the model, or theory, to grow. Good models should have elements that directly correspond to objects in the real world. They should provide a reliable answer to our question(s).

The incentives to model a system's behavior without checking model results against reality are frequently high. If the discipline is sufficiently well established, and if its proponents are powerful and enjoy high social status, their models will be well received. If all those models make predictions ultimately shown to be wrong, and yours does too, you are accepted in the discipline. If you do better in explaining the system's behavior, it may be regarded as luck. However, if your model fails to be better than others, you are in trouble. Therefore, it may be more convenient for you to stay with the majority, not challenging the dominant paradigm or school of thought. But be aware, others may come and challenge you.

In the modeling process that we propose here, it is relatively easy to change an assumption and determine the resulting effect on system-wide behavior. It is therefore possible for you to challenge established paradigms—if the results are closer to reality, you may be on your way to a new paradigm.

1.1.2   Static, Comparative Static, and Dynamic Models

Most models fit in one of three general classes. To first type belong models that represent a particular phenomenon at a point of time, a static model. For example, a map of the United States depicts the location and size of a city or the rate of infection with a particular disease, each in a given year. The second class comprises comparative static models that compare some phenomena at different points in time. This is like using a series of snapshots to make inferences about the system's path from one point in time to another without modeling that process itself.

Other models describe and analyze the very processes underlying a particular phenomenon. An example of this would be a mathematical model that describes the demand for and supply of a good as a function of its price. Choosing a static modeling approach, we may want to find the price under which demand and supply are in equilibrium and investigate the properties of this equilibrium: Is the equilibrium price a unique one or are there other prices that balance demand and supply? Is the equilibrium stable or are small perturbations to the system accompanied by a movement away from the equilibrium? Such equilibrium analysis is widespread in economics.

Alternatively, a model could be developed to show the *changes* in demand and supply *over time*. Dynamic models are those that try to reflect changes in real or simulated time and take into account that the model components are constantly evolving as a result of previous actions.

With the advent of easy-to-use computers and software, we can all build on the existing descriptions of a system and carry them further. The world is not a static or comparative static process, and so the models treating it in that way will become obsolete and are perhaps even misleading. We can now investigate in great detail and with great precision the system's behavior over time, its movement toward, or away from, equilibrium positions, rather than restricting the analysis to the equilibrium itself.

Throughout the social, biological, and physical sciences, researchers examine complex and changing interrelationships among factors in a variety of disciplines. What are the impacts of a change in the El Niño winds, not only on weather patterns, but also on the cost of beef in the United States? How does the value of the Mexican peso affect the rate of oil exploration in Alaska? Every day, scientists ask questions like these that involve dissimilar frames of reference and even disciplines. This is why we must understand the dynamics and complex interrelationships among diverse systems in our increasingly complicated world. A good set of questions is the start—and often the conclusion—of a good model. Such questions help the researcher remain focused on the model, without becoming distracted by the myriad of random details that surround it.

Through computer modeling we can study processes in the real world by sketching simplified versions of the forces assumed to underlie them. As an example, you might hypothesize that cities drew workers from farmlands as both became greater users of technology, causing a surplus of jobs in the city and a surplus of labor in the countryside. Another factor could be the feasibility of moving from an agricultural area to the city. A basic version of this model might abstract away from many of the factors that encourage or discourage such migration, in addition to those directly related to job location and the feasibility of relocation. This model could leave behind a sufficiently good predictor of migration rates, or it might not. If the model does, you can reexamine it. Did your abstractions eliminate an important factor? Were all your assumptions valid? You can revise your model, based on the answers to these questions. Then, you can test the revised model for its predictions. You should now have an improved model of the system you are studying. Even better, your understanding of that system will have grown.

You can better determine whether you asked the right questions, included all the important factors in that system, and represented those factors properly.

Elementary to modeling is the idea that the model should be kept simple, even simpler than the cause and effect relationship it studies. We add complexities to the model only when it does not produce the real effects. Models are sketches of real systems and are not designed to show all the system's many facets. Models aid us in understanding complicated systems by simplifying them; such multifaceted models would defeat this purpose.

Models study cause and effect; they are causal. The modeler specifies initial conditions and relations among these elements. The model then describes how each condition will change in response to changes in the others. In the example of farm workers moving to the city, workers moved in response to a lack of jobs in the countryside. But, more workers in the city would raise the demand for food in city markets, raising the demand for farm labor. Thus, the demand for labor between the city and the country would shift, leading to migration in both directions and changes in migration over time.

The initial conditions selected by the modeler could be actual measurements (the number of people in a city) or estimates (how many people will be there in 4 years, given normal birth rate and other specified conditions). Such estimates are designed to reflect the process under study, not provide precise information about it. Therefore, the estimates could be based on real data or be the reasonable guesses of a modeler, based on experience with the process. At each step in the modeling process, documentation of the initial conditions, choice of parameters, presumed relationships, and any other assumptions is always necessary, especially when the model is based on the modeler's guesses.

1.1.3 Model Components

Model building begins, of course, with the properly focused question. Then we must decide on the boundaries of the system that contains the question. We must choose the appropriate time step and the level of needed detail. But these are verbal descriptions. Sooner or later we must get down to the business of actually building the model. The very first step in that process is the identification of the state variables, those variables that will indicate the status of our system through time. These variables carry the knowledge of the system from step to step throughout the run of the model—they are the basis for the calculation of all the rest of the variables in the model.

Generally, there are two kinds of state variables, conserved and nonconserved. Conserved variables are such things as the population in a country or the water behind a dam. They do not take on negative values. Nonconserved state variables are temperature or price, for example, and they might be able to take on negative values (such as temperature measured on a Celsius or Fahrenheit scale) or they might not (price).

Control variables are the ones that directly change the state variables. They can increase or decrease the state variables through time. Birth (per time period) or water inflow (to a reservoir) or heat flow from a hot body all are examples of controls affecting state variables.

Transforming or converting variables are sources of information used to change the control variables. Such a variable might be the result of an equation based on still other transforming variables or parameters. The birth rate, the evaporation rate, and the heat loss coefficient are examples of transforming variables.

The components of a model are expected to interact with each other. Such interactions engender feedback processes. *Feedback* describes the process wherein one component of the model initiates changes in other components and those modifications lead to further changes in the component that set the process in motion.

Feedback is said to be negative when the modification in a component leads other components to respond by counteracting that change. As an example, the increase in the need for food in the city caused by workers migrating to the city leads to a demand for more laborers in the farmlands. Negative feedback is often the engine that drives supply and demand cycles toward some equilibrium. The word *negative* does not imply a value judgment—it merely indicates that feedback tends to negate an initial stimulus for change.

In positive feedback, the original modification leads to changes that reinforce the stimulus that started the process. For example, if you feel good about yourself and think you are doing well in your work, you may work harder. Because you work hard, you are more successful, which makes you feel good about yourself and enhances your chances of doing well in your work. On the other hand, if you feel bad about yourself and how you are doing in your work, you may lose motivation and not work very hard. But because you did not work hard enough, your performance and progress will decline, making you feel bad about yourself. As another example, the migration of farm workers to a city attracts more manufacturers to open plants in that city, which attracts even more workers. Another economic example of positive feedback was noticed by Brian Arthur—firms that are first to occupy a geographic space will be the first to meet the demand in a region and are the most likely to build additional branch plants or otherwise extend their operations. The same appears to be true with pioneer farmers—the largest farmers today were among the first to cultivate the land.

Negative feedback processes tend to counteract a disturbance and lead systems back toward an equilibrium or steady state. One possible outcome of market mechanisms would be that demand and supply balance each other or fluctuate around an equilibrium point, due to lagged adjustments in the productive or consumptive sector. In contrast, positive feedback processes tend to amplify any disturbance, leading systems away from equilibrium. This movement away from equilibrium is apparent in the example of how you feel about yourself affects your performance.

People from different disciplines perceive or interpret the role and strength of feedback processes differently. Neoclassical economic theory, for example, is typically preoccupied with market forces that lead to equilibrium in the system.

Therefore, the models are dominated by negative feedback mechanisms, such as price increases in response to increased demand. The work of ecologists and biologists, in contrast, is frequently concerned with positive feedbacks, such as those leading to insect outbreaks or the dominance of hereditary traits in a population.

Most systems contain both positive and negative feedback; these processes are different and vary in strength. For example, as more people are born in a rural area, the population may grow faster (positive feedback). However, as the limits of available arable land are reached by agriculture, the birth rate slows, at first perhaps for psychological reasons but eventually for reasons of starvation (negative feedback).

Nonlinear relationships complicate the study of feedback processes. An example of such a nonlinear relationship would occur when a control variable does not increase in direct proportion to another variable but changes in a nonlinear way. Nonlinear feedback processes can cause systems to exhibit complex—even chaotic—behavior.

A variety of feedback processes engender complex system behavior, some of which we will encounter in this book. Let us develop a simple model to illustrate the concepts of state variables, flows, and feedback processes. We will then return to discuss some "principles of modeling" that will help you to develop the model building process in a set of steps.

1.1.4   Modeling in STELLA

For a save-disabled copy of STELLA and the models in this book, go to the following Internet site: www.iseesystems.com/ModelingEconomicSystems

To explore modeling with STELLA, we will develop a basic model of the dynamics of a fish population. Assume you are the sole owner of a pond that is stocked with 200 fish that all reproduce at a fixed rate of 5% per year. For simplicity, assume also that none of the fish die. How many fish will you own after 20 years?

In building the model, we will utilize all four of the graphical "tools" for programming in STELLA. On opening STELLA, you will be faced with the "Map" layer, which we will not need now. It is available to lay out the structure of a model without quantifying initial conditions or interrelationships among model elements. To get to the "Model" layer—that part of the software in which we develop and specify the actual model—click on the Model tab on the left-hand side of the frame.

Fig. 1.1 Main building
blocks for model
development

Fig. 1.2 State variable of
the fish dynamics model

FISH

The Model layer allows you to specify your model's initial conditions and functional relationships. To do so will require use of symbols, or "building blocks," for stocks, flows, converters, and connectors (information arrows), shown, respectively, from left to right in Fig. 1.1.

We begin with the first tool, a stock (rectangle). In our example model, the stock will represent the number of fish in your pond. Click on the rectangle with your mouse, drag it to the center of the screen, and click again. Type in FISH and you should get what's shown in Fig. 1.2

This is the first state variable in our model. Here we indicate and document a state or condition of the system. In STELLA, this stock is known as a *reservoir*. In our model, this stock represents the number of fish of the species we are studying that populate the pond. If we assume that the pond is one square kilometer large, the value of the state variable FISH is also its *density*, which will be updated and stored in the computer's memory at every step of time (DT) throughout the duration of the model. The fish population is a stock, something that can be contained and conserved in the *reservoir*; density is not a stock, it is not conserved. Nonetheless, both of these variables are state variables. So, because we are studying a species of fish in a *specific area* (one square kilometer), the population size and density are represented by the same rectangle.

Inside the rectangle is a question mark. This is to remind us that we need an initial or starting value for all state variables. Double-click on the rectangle. A dialogue box will appear. The box is asking for an initial value. Add the initial value you choose, such as 200, using the keyboard or the mouse and the dialogue keypad. When you have finished, click on OK to close the dialogue box. The question mark will have disappeared.

We must decide next what factors control (that is, add to or subtract from) the number of fish in the population. Because we assumed that the fish in your pond never die, we have one control variable: REPRODUCTION. We use the *flow* tool (the right-pointing arrow, second from the left) to represent the control variable, so named because they control the states (variables). Click on the flow symbol, then click on a point about 2 in. to the left of the rectangle (stock) and drag the arrow to POPULATION, until the stock becomes dashed, and release. Label the circle REPRODUCTION. Figure 1.3 shows what you will have.

Fig. 1.3 Stock with outflow

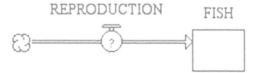

REPRODUCTION   FISH

Here, the arrow points only into the stock, which indicates an inflow. But, you can get the arrow to point both ways if you want it to. You do this by double-clicking on the circle in the flow symbol and choosing Biflow. (Conversely, if your flow is drawn as a biflow and you only want it to go one direction, check the Uniflow box.) Biflow enables you to add to the stock if the flow generates a positive number and subtract from the stock if the flow is negative. In our model, of course, the flow REPRO-DUCTION is always positive and newly born fish go only *into* the population. Our control variable REPRODUCTION is a uniflow: "new fish per annum."

Next we need to know how the fish in our population reproduce. Not the biological details, just how to accurately estimate the number of new fish per annum. One way to do this is to look up the birth rate for the fish species in your pond. Say we find that the birth rate = 5 new fish per 100 adults each year, which can be represented as a transforming variable. A transforming variable is expressed as a *converter*, the circle that is second from the right in the STELLA toolbox. So far REPRODUCTION RATE is a constant, later we will allow the reproduction rate to vary. The same clicking and dragging technique that got the stock on the screen will bring up the circle. Open the converter and enter the number of 0.05 (5/100). Down the side of the dialogue box is an impressive list of "built-in" functions that we can use for more elaborate models.

At the right of the STELLA toolbox is the *connector* (information arrow). We use the connector to pass on information (about the state, control, or transforming variable) to a circle, to control or transforming variable. In this case, we want to pass on information about the REPRODUCTION RATE to REPRODUCTION. Once you draw the information arrow from the transforming variable REPRODUC-TION RATE to the control and from the stock FISH to the control, open the control by double-clicking on it. Recognize that REPRODUCTION RATE and FISH are two required inputs for the specification of REPRODUCTION. Note also that STELLA asks you to specify the control: REPRODUCTION = ... "Place right-hand side of equation here"

Click on REPRODUCTION, then on the multiplication sign in the dialogue box, and then on FISH to generate the equation

$$\text{REPRODUCTION} = \text{REPRODUCTION RATE} * \text{FISH}. \qquad (1.1)$$

Check the box "Non-negative" and then click on OK and the question mark in the control REPRODUCTION disappears. Your STELLA diagram should now look as shown in Fig. 1.4.

Next, we set the temporal (time) parameters of the model. These are DT (the time step over which the stock variables are updated) and the total time length of a

Fig. 1.4 Basic population
dynamics model

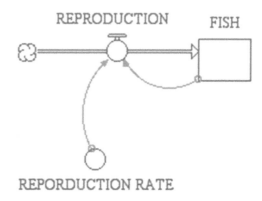

REPRODUCTION   FISH

REPORDUCTION RATE

Fig. 1.5 Icons to display
model output

model run. Go to the RUN pull-down menu on the menu bar and select Time
Specs. . . A dialogue box will appear in which you can specify, among other things,
the length of the simulation, the DT, and the units of time. We arbitrarily choose
DT = 1, length of time = 20, and units of time = years.

To display the results of our model, click on the graph icon and drag it to the
diagram. If we wanted to, we could display these results in a table by choosing the
table icon instead. The STELLA icons for graphs and tables are shown in Fig. 1.5.

When you create a new graph pad, it will open automatically. To open a pad
that had been created previously, just double-click on it to display the list of
stocks, flows, and parameters for our model. Each one can be plotted. Select FISH
to be plotted and, with the ≫ arrow, add it to the list of selected items. Then set the
scale from 0 to 600 and check OK. You can set the scale by clicking once on the
variable whose scale you wish to set and then on the ↕ arrow next to it. Now you can
select the minimum on the graph, and the maximum value will define the highest
point on the graph. Rerunning the model under alternative parameter settings will
lead to graphs that are plotted over different ranges. Sometimes these are a bit
difficult to compare with previous runs, because the scaling has changed.

Would you like to see the results of our model so far? We can run the model by
selecting RUN from the pull-down menu. We get results shown in Fig. 1.6.

We see a graph of exponential growth of the fish population in your pond. This is
what we should have expected. It is important to state beforehand what results you
expect from running a model. Such speculation builds your insight into system
behavior and helps you anticipate (and correct) programming errors. When the
results do not meet your expectations, something is wrong and you must fix it.
The error may be either in your STELLA program or your understanding of the
system that you wish to model—or both.

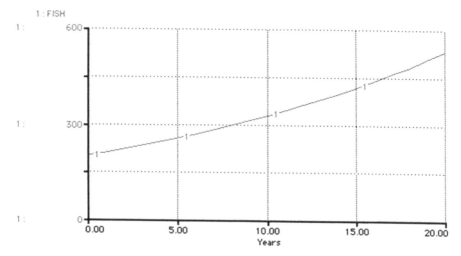

Fig. 1.6 Results of the basic population model

What do we really have here? How does STELLA determine the time path of our state variable? Actually, it is not very difficult. At the beginning of each time period, starting with time = 0 years (the initial period), STELLA looks at all the components for the required calculations. The values of the state variables will probably form the basis for these calculations. Only the variable REPRODUCTION depends on the state variable FISH. To estimate the value of REPRODUCTION after the first time period, STELLA multiplies 0.05 by the value FISH (@ time = 0) or 200 (provided by the information arrows) to arrive at 10. From time = 1 to time = 2, the next DT, STELLA repeats the process and continues through the length of the model. When you plot your model results in a table, you find that, for our simple fish model, STELLA calculates fractions of fish from time = 1 onward. This problem is easy to solve; for example, by having STELLA round the calculated number of fish—there is a built-in function that can do that—or just by reinterpreting the population size as "thousands of fish."

This process of calculating stocks from flows highlights the important role played by the state variable. The computer carries that information—and only that information—from one DT to the next, which is why it is defined as the variable that represents the *condition* of the system.

You can drill down in the STELLA model to see the parameters and equations that you have specified and how STELLA makes use of them. Click on the Equation tab on the left side of your screen and you will see the following listing:

```
FISH(t) = FISH(t - dt) + (REPRODUCTION) * dt
INIT FISH = 200
INFLOWS:
REPRODUCTION = REPORDUCTION_RATE*FISH
REPORDUCTION_RATE = .05
```

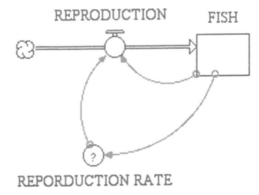

Fig. 1.7 Basic population model with endogenous reproduction rate

Note how the fish population in time period t is calculated from the population one small time step, DT, earlier and all the flows that occurred during a DT.

The model of the fish population dynamics is simple. So simple, in fact, we could have solved it with pencil and paper, using analytic or symbolic techniques. The model is also linear and unrealistic. So let us add a dimension of reality—and explore some of STELLA's flexibility. This may be justified by the observation that, as populations get large, mechanisms set in that influence the rate of reproduction.

To account for feedback between the size of the fish population and its rate of reproduction, an information arrow is needed to connect FISH with REPRODUC-TION RATE. The connection will cause a question mark to appear in the symbol for REPRODUCTION RATE (Fig. 1.7). The previous specification is no longer correct; it now requires FISH as an input.

Open REPRODUCTION RATE. Click on the required input FISH. The relation-ship between REPRODUCTION RATE and FISH must be specified in mathe-matical terms, or at least, we must make an educated guess about it. Our educated guess about the relationship between two variables can be expressed by plotting a graph that reflects the anticipated effect one variable (REPRODUCTION) will have on another (FISH). The feature we will use is called a *graphical function*.

To use a graph to delineate the extended relationship between REPRODUC-TION RATE and FISH, we click on Become Graph. Set the limits on the FISH at 2 and 500; set the corresponding limits on the REPRODUCTION RATE at 0 and 0.20, to represent a change in the birth rate when the population is between 0 and 500. Here we are using arbitrary numbers for a made-up model. Finally, use the mouse arrow to draw a curve from the maximum birth rate and population of 2 to the point of zero birth rate and population of 500.

Suppose a census were taken at three points in time. The curve we just drew goes through all three points. We can assume that, if a census had been taken at other times, it would show a gradual transition through all the points. This sketch is good enough for now (Fig. 1.8). Click on OK.

Before we run the model again, let us speculate what our results will be. Think of the graph for FISH through time. Generally, it should rise, but not in a straight line.

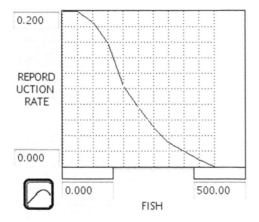

Fig. 1.8 Graphically defined reproduction rate

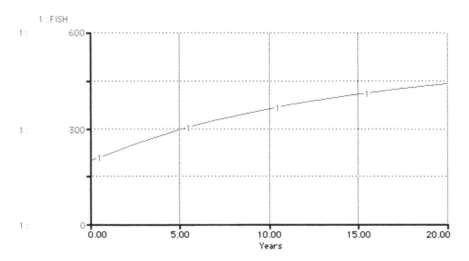

Fig. 1.9 Model results of the basic population model with endogenous reproduction rate

At first the rise should be steep: the initial population is only 200, so the initial birth rate should be very high. Later it will slow down. Then, the population should level off at 500, when the population's density would be so great that new births tend to cease. Run the model. See Fig. 1.8 for the model results. We were right (Fig. 1.9).

This problem has no analytic solution, only a numerical one. We can continue to study the sensitivity of the answer to changes in the graph and the size of DT. We are not limited to a DT of 1. Generally speaking, a smaller DT leads to more accurate numerical calculations for updating state variables and, therefore, a more accurate answer. Choose Time Specs from the RUN menu to change the DT.

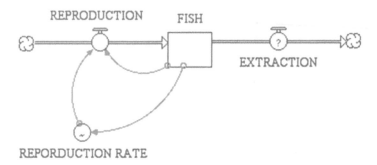

Fig. 1.10 Basic population model with extraction

Change the DT to reflect ever-smaller periods until the change in the critical variable is within measuring tolerances. We may also change the numerical technique used to solve the model equations. Euler's method is chosen as a default. Two other methods, Runge–Kutta-2 and Runge–Kutta-4, are available that update state variables in different ways. We will discuss these methods later.

Start with a simple model and keep it simple, especially at first. Whenever possible, compare your results against measured values. Complicate your model only when your results do not predict the available experimental data with sufficient accuracy or when your model does not yet include all the features of the real system that you wish to capture. For example, as the owner of a pond, you may want to extract fish for sale. Assume the price per fish is $5 and constant. What are your revenues each year if you wish to extract fish at a constant rate of 3% per year? To find the answer to this question, define an outflow from the stock FISH. Click on the converter, then click onto the stock to have the converter connected to the stock, and then drag the flow from the stock to the right. Now fish disappear from the stock into a "cloud." We are not explicitly modeling where they go. Figure 1.10 shows what you should have developed thus far as your STELLA model.

Next, define a new transforming variable called EXTRACTION RATE and set it to 0.03. Specify the outflow as

$$\text{EXTRACTION} = \text{EXTRACTION RATE} * \text{FISH} \tag{1.2}$$

after making the appropriate connections with information arrows. Then create two more transforming variables, one called PRICE, set it to 5, and the other called REVENUES, specify it as

$$\text{REVENUES} = \text{EXTRACTION} * \text{PRICE} \tag{1.3}$$

Your model should look now as in Fig. 1.11.

Double-click on the graph pad and select REVENUES to plot the fish stock and revenues in the same graph. The time profile of the revenue streams generated by your pond is shown in Fig. 1.12.

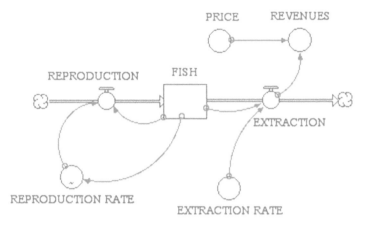

Fig. 1.11 Completed fish model

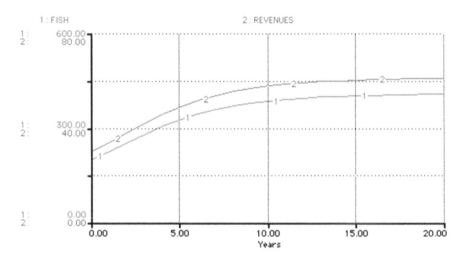

Fig. 1.12 Results from completed fish model

You can easily expand this model; for example, to capture price changes over time, unforeseen outbreaks of diseases in your pond, or other problems that may occur in a real-world setting. When your model becomes increasingly complicated, try to keep your STELLA diagram as organized as possible, so it clearly shows the interrelationships among the model parts. A strong point of STELLA is its ability to demonstrate complicated models visually. Click on the arrow symbol (Fig. 1.13) to move model parts around the diagram; use the "paintbrush" to change the color of icons. The "dynamite" will blast off any unnecessary parts of the model.

Be careful when you blast away connectors (information arrows). Move the dynamite to the place at which the information arrow is attached and click on that spot. If you click, for example, on the convertor itself, it will disappear, together with the connector.

Fig. 1.13 STELLA tools to
keep models organized

Fig. 1.14 Sector symbol

Fig. 1.15 Text symbol for
model annotation

As the model grows, it will contain an increasing number of submodels, or modules. You may want to protect some of these modules from being changed. To do this, click on the "Sector" symbol (Fig. 1.14) and drag it over that module or modules you want to protect. To run the individual sectors, go to Sector Specs... in the RUN pull-down menu and select the ones that you wish to run. The values of the variables in the other sectors remain unaffected.

By annotating the model, you can remind yourself and inform others of the assumptions underlying your model and its submodels. This is important in any model, but especially in larger and more complicated models. To do this, click on the "Text" symbol (the letter T) and drag it into the diagram. Then type in your annotation (Fig. 1.15).

The tools we mentioned here are likely to prove very useful when you develop more complicated models and when you want to share your models and their results with others. STELLA contains very helpful tools, which we hope you will use extensively. Space limitations preclude us from describing all of STELLA's features. You will probably want to explore on your own many such features—a little experimentation or browsing through STELLA's help features will go a long way.

Make thorough use of your model, running it over again and always checking your expectations against its results. Change the initial conditions and try running the model to its extremes. At some point you will want to perform a formal sensitivity analysis. Later, we will discuss STELLA's excellent sensitivity analysis procedures.

1.1.5   Modeling Principles

Although the title of this section might seem somewhat ostentatious, we surely have learned something general about modeling after years of building and running models. These steps are a distillation of the process. The steps are not sacred: they are intended as a guide to start you in the process. As you proceed in model building, return to the list, challenge, and refine the principles.

1. Identify the problem and what you want to achieve in the model. Pose the questions that the model is to answer. Break down very large or complex problems into subsystems and identify what you hope to achieve in modeling each subsystem. Ask yourself if the problem is descriptive or predictive.

2. Define your state variables and do not forget to note the units of measurement. You can type these units in the dialogue box, using braces, { and }, to separate them from the specification of state variables. STELLA will read the text that you write in braces as comments. You may also want to use STELLA's built in unit editor, which you can access through the "Units..." button on the right-hand side of the dialog box that you see after you double-click on a stock or convertor. At this stage, try to keep your model as uncomplicated as possible. Purposely shun complexity at the start.

3. Stipulate your control variables, the flow controls into and out of the state variables, and specify their units. List the functional relationships between the state variables and control variables. For each control variable, ask which state variables are its donors and which are its recipients. Focus on the most essential features. Use one type of control to depict a class of similar controls. You will add the others in step 10.

4. Determine the parameters and units for your control variables. What is the function of these controls and their parameters?

5. Look at your model. Does it violate any laws of physics, economics, or other discipline? For instance, have you allowed for the conservation of mass or energy (or whatever is relevant to your model)? Are the units consistent? You may have to use conditional statements to avoid such violations as division by zero or negative volumes or prices. Fully document your parameters, initial values, assumptions, and equations before going on.

6. For an idea of how the model will work, select time and space parameters over which to examine its dynamic behavior. Choose the length of each time interval (DT) for which state variables are to be updated by reference to the space over which dynamics occur; mainly, by reference to the fastest rate of change you expect in your model. Then choose the numerical computation procedure (Euler's method, Runge–Kutta-2, or Runge–Kutta-4) to calculate flows. Prepare a graph of the most important variables and guess their variation before running the model.

7. Run the model. Was the graph you prepared reasonable? Experiment with other time intervals over which to update your state variables. Experiment with other integration techniques.

8. Take your parameters to their reasonable extremes. Do they still make sense? Rework the model to eliminate errors and anomalies.

9. Examine your results. How do they compare to the experimental data? To do this, you may have to mimic a lab experiment by shutting off parts of your model.

10. Using the understanding gained by going through these steps, reassess your parameters, even your model structure, to allow for greater complexity and include the exceptions to your experimental results. Repeat steps 1–10. Finally, pose a new set of interesting questions, derived from your experience with the model.

It is not necessary to apply these steps in exactly the order given here. Develop your own model and improve your modeling skills. Return to the list periodically to see if you find these steps practical, inclusive, and logical.

Keep in mind that modeling can have four purposes. First, models enable you to experiment: A good model of a system lets you alter its components and experience the effect of such changes on the system. Second, good models give you insight into the future course of a dynamic system. Third, good models lead you to further questions about a system, what underlies its behavior, and how applicable are the principles discovered in the modeling process to other systems. Fourth, good models are good thought-organizing devices.

1.1.6   Model Confirmation

When do you know that you developed an appropriate model of real events? Giving an answer to this question often is rather difficult. By definition, all models abstract away from some aspects of reality that the modeler perceives less relevant than others—less relevant, that is, with respect to the questions posed at the beginning of the modeling process. As a result, the model is a product of the modeler's perception. Consequently, one model is likely not the same as the models developed for the same system by other modelers, who may have their own, individual perceptions and questions. Plurality of, and competition among, models is therefore required to improve our collective understanding of real-world processes. The more open and flexible is the modeling approach, and the more people engage in modeling, the larger the diversity that can foster new knowledge.

By enclosing a selected number of system components in the model and determining the model–system's behavior over time solely in response to the forces inside the model, the model becomes closed. Real systems, in contrast, are not closed but open, allowing for new, even unprecedented development in response to highly infrequent but dramatic changes in their environment. It is therefore not possible to truly or completely "verify" a model by comparing model results to the behavior of the real system. Extenuating circumstances may have led the real system behave differently from the model; even though the model itself may not have been incorrect, it may have been just incomplete with regard to those circumstances. Such circumstances will always be present, precisely because the model's goal is to capture only the "essentials" of the real system and abstract away from other factors. Consequently, verification of a model can be done only with regard to the consistency, or logical accuracy, of its internal structure.

Generally, there are two ways to test a model. First, one can withhold some of the basic data that was used to set up the model, data that represent the real world to the extent that it can be measured. Then the model can be used to "predict" the data used. For example, one might develop a model to predict the future population of the US based on actual data from, say, 1900 to 1960 and then use the resulting model to predict the (known) population data from 1970 to 1990, by estimating the birth and death rates for each cohort between 1970 and 1990. If the "prediction" is a success, the later actual data can be incorporated into the model, and a real prediction for the next 20 or 30 years can be made with an improved degree of confidence. Another way to test the "goodness" of a model is to predict the condition in some heretofore unmeasured arena and then proceed to measure these variables in the field. For example, suppose that a model is being set up to predict the location of an endangered species. The model is built and calibrated on the known habitats and then applied to the rest of the likely geography to qualify these places as likely locations. A good example of this is the effort to model ecosystems. If we want to establish a quasi-stable base for the determination of various forms of human impact, we need to establish the natural dynamics of the system before today's impacts began. We would then be describing a system that we have never seen and have no data on, and yet, our model may represent the best possible description of it.

Some of our system modeling is not for the purposes of explanation or prediction at all. In some cases, we model a system to "retrieve" its original but unmeasurable conditions. We do this to provide a reasonable reference system to which ensuing states of the system can be compared or referenced. For example, when trying to determine the current degree of distress in an ecosystem, we dynamically model all of the natural elements that we can find or imagine in the original system, without the known human disturbances. The long-term conditions in the model are taken to be the presettlement condition of that ecosystem.

If it is not possible, by definition, to verify a model by comparing its results to the performance of the real system, how can we know that we really captured the essentials of that system in the model? More than one model of the same system can produce the same result. We may compare the model results to reality, not to verify but to confirm, and if we are unable to reproduce at least the trends observed in the real system, we know something is missing or wrong, and we are forced to revise our model or check the accuracy of the data that went into specifying the model parameters and initial conditions. We can invoke the "reasonability" test. We can ask if the results are reasonable insofar as our experience is concerned. We can push the model to its extremes and ask again how reasonable the results are. Do the results make sense to us? Do the results fit with the known physical laws, the laws of conservation where they apply, and so forth? This is why it is so important to speculate on what the results should be before you run the model. Not only does such speculation help you avoid accepting blindly the model results, but it also sharpens and builds your sense of how systems operate dynamically.

Ironically, things also may become more problematic if the model results coincide well with our observations of the real system. The problem, for example, lies in the possibility that errors in the model cancel each other. Such misspecifications are difficult to detect, and this is why many chapters of this book will start with a theory of model behavior rather than with casual observations of real systems. Combining theory, observations, and indeed, intuition in a disciplined way in dynamic models is especially important when we use easy-to-learn, easy-to-use software packages that enable us to develop models with which we could get ahead of ourselves. At each point in the model construction process it is important to be able to justify the assumptions that are made.

Once the model is built on a theoretical base and observations, or "reasonable" initial conditions and parameter values, we let it yield the consequences of the forces built into the model. The choice of observations vs. "reasonable" values is basically a choice of providing a predictive or descriptive model. For a description of the role of feedback mechanisms for system development, it is frequently sufficient to concentrate on those forces and the appropriate parameter range rather than precise numerical values.

To confirm our model results we may compare them to observational data. The greater is the number of instances in which model results and reality coincide under a variety of different scenarios, the more probable it is that the model captures the essential features of the real system it attempts to portray. We can increase our confidence in the model further by comparing its results to other models for the same or similar systems. The latter approach is of particular interest if the model was descriptive rather than predictive. Finally, we hope that we develop, through practice in modeling, experience and new understanding of system dynamics that enable us to more easily detect problems in model specifications. This is a learning process; we hope our book makes its contribution.

1.1.7   Modeling of Natural Resource Use

A principal goal of modeling the dynamics of natural resource use in economics is to develop quantitative and integrated models that effectively deal with the inherent uncertainty in these very complex systems. For more than 3 decades, dynamic models have been used to assess the adequacy of natural resources to meet human needs. Many models focus on the relative strengths of positive and negative feedback processes that influence the availability of natural resources. Most notable is the question whether technical change, the use of substitutes, or recycling can compensate for a decrease in the quality or quantity of an essential resource. As extraction proceeds, resource quality and quantity tend to decrease, and thus, cost of extraction may increase. Recycling, use of substitutes, and technical change, on the other hand, may drive down the unit cost of extraction, opposing the forces of resource depletion. The relative strength and timing of these forces are difficult to assess a priori because of the many interconnections and feedbacks among depletion, recycling, substitution, and technical change. Most economists will admit that their models do not include what they necessarily call *externalities*. This open-ended form of dynamic modeling enables us to rather readily incorporate such externalities and determine their importance. Modeling the dynamics of these systems with the computer may improve our understanding of these systems, and may help us identify crucial relationships that determine the systems' dynamics.

With the popularization of computer models for natural resource policy-making, attention is now directed at a number of closely related issues that potentially dilute the effectiveness of those models in decision-making (Smil 1993). First, all models are abstractions from reality and, therefore, incomplete. It is inappropriate to think of models as anything but crude, although in many cases absolutely essential, abstract representations of reality. What to include or exclude from a model depends on the goal of the model and is a choice that is ultimately made within a cultural, socioeconomic, and scientific context. As a result of the diversity in perceptions and aspirations of modelers with different cultural and scientific backgrounds, there are a large number of different methods for modeling resource use in economic systems, each of which has a unique set of strengths and weaknesses.

Second, unlike models of physical systems that are based on rather well-established rules, models of human systems are prone to misspecifications of the complexities of human behavior. The context dependence of human activities together with the variability in rules used to make decisions limits the applicability of models for descriptive and predictive purposes. The predictive ability of models is further reduced by the possibility of the occurrence of previously unencountered constraints on, and development possibilities for, the systems. The latter is frequently subsumed under the notion of novelty and most prominent in the case of technological breakthroughs or discoveries of new deposits of a resource.

Third, models of natural resource use tend to become larger and more complex as modelers and decision-makers ask to include aspects of the system that have previously been treated as exogenous to the model. There is an increased tendency to "close" models; that is, to minimize the number of exogenously specified parameters and relationships. However, with increasing scale and model complexity, only experts, frequently only the creators of the models themselves, can interpret the structure, dynamics, and results of the model.

Fourth, complex systems are characterized by substantial uncertainty that must be addressed as a central part of the model building and evaluation process, not relegated to the sidelines as is commonly the case. The task is to manage the uncertainty of each model in order to gain the highest quality information from it. The enormous uncertainty in the relationship between the economy and the environment calls for a plurality of modeling approaches and perspectives, hence our continual haranguing of the reader to keep the models simple.

Given the number and complexity of models developed to assess issues of natural resource use, decision-makers frequently are faced with the following dilemma. If they choose one model over another, decision-makers effectively transfer part of their power to the modeler. By using a number of models on which to base their decisions, the nonexpert policy-makers are left to reconcile the different assumptions underlying the models, thus opening themselves up for criticism irrespective of the value of their decision. Many modeling exercises are full of hyper-precise numerical data, specious sophistication about uncertainties, and unsupported hunches and loaded rhetoric, leaving policy-makers with little hope of sorting through the various approaches.

Not surprisingly, the call for more transparent, flexible, yet more comprehensive models is common (Meadows and Robinson 1985). With this book, we wish to promote a revolution in systems thinking and hope that we further the democratization of the modeling process. Because the purpose of this book is to teach systems thinking rather than data collection and analysis, the majority of the models presented here do not make use of real-world data. Many of the concepts developed here, however, are being applied to real-world systems. We challenge you to carry the models of this book further and contribute to the understanding of these systems.

1.1.8   Extending the Modeling Approach

Unlike most of today's computer languages, STELLA lets us focus on exploring the features of a dynamic system, rather than writing a program to follow some complicated, unintuitive syntax. With its easy-to-learn and easy-to-use approach to modeling, STELLA provides us with a number of features that enhance the development of modeling skills and collaboration of modelers. First there is the knowledge-capturing aspect of STELLA. The way we employ STELLA leads not only to a program that captures essential features of a system, but more important, results in a process that involves the assembly of experts who contribute their specialized knowledge to a model of a system. Experts can see the way in which their knowledge is incorporated into the model because they can pick up the fundamental aspects of STELLA very quickly. They can see how their particular part of the whole is performing and judge what changes may be needed. They can see how their part of the model interacts with others and how other specialists formulate their own contributions. As a result, these experts are likely to be more confident in the whole model and its accuracy than if they had entrusted a programmer with their insight into the system's workings. Once engaged in this modeling process, experts will sing its praises to other scientists. This is the knowledge-capturing aspect of the modeling process.

These same experts are likely to take STELLA into the depths of their own discipline. But, most important, they will return to their original models and repair them to meet broader challenges than first intended. Thus models grow along with the expertise and understanding of the experts. This is the expert-capturing part of our process. It is based not only on informed consensus but it has the possibility of continuing growth of the central model.

So we avoid the cult of the central modeler, working on a mainframe computer, who claims to have understood and captured the meaning of the experts. We avoid producing a group of scientists who are unsure of the degree to which their knowledge has been captured and so, in their conservative default position, eventually deny any utility or even connection to the model and generally retain a negative opinion of this and perhaps even all such models.

The process is not risk free, and it requires delicate organization. The scientists who participate in the modeling endeavor do reveal their approach to their field. That approach may contain errors and omissions, and thus, they may be prone to increased criticism. However, these scientists have to be congratulated for their courage to open up new areas for scientific endeavors and for opening themselves up to criticism. The process does, however, promise to make young scientists wiser and older scientists younger.

A general strategy for modeling with experts is to build a STELLA model of the phenomena that all expect will be needed to answer the questions posed at the outset. That model should cover a space and have a time step commensurate with the detail needed for the questions. The whole space and time needed for the model may exceed reasonable use of the desktop computer. Eventually, any modeling enterprise may become so large that the program STELLA is too cumbersome to use. However, translators exist for the final STELLA model, converting the STELLA equations into either C+ or FORTRAN code. Translated models can be duplicated and put into adjoining cells on a landscape and run simultaneously on large computers.

The various modeling processes are now being combined into a seamless one, running from desktop to supercomputer and back, with little special effort on the part of the body of experts. As an example, in an ecological model studying sage grouse (Westervelt and Hannon 1993) in the desert, we used STELLA to combine the expertise of a variety of life science professionals. The STELLA generic model could be electronically translated into C^+ or FORTRAN and applied to a series of connected cells; in this example, as many as 120,000 such cells. These now cellularized models were then electronically initialized with appropriate maps from a specific Geographic Information System. The model could then be run on a large parallel processing computer or a network of paralleled workstations. Thus, we could connect STELLA's knowledge-capturing features to the most powerful computers available.

The interdisciplinary nature of landscape ecology and environmental management has led researchers in these fields to make good use of a cellular approach to building dynamic spatial models in STELLA and to running these models on ever more powerful computers. Ruth and Pieper (1994) used a different, but similar model to study the spatial dynamics of a rising sea level. Their model uses a relatively small set of interconnected cells to describe the physical processes of coastal erosion and sedimentation. Each cell of the model held site-specific data. The model was run to cover a small part of the total area. Then the cells were moved across the landscape to cover part of the area that was not covered previously by the model. By rerunning and "moving" the cells over and over again, a large area could be covered with a relatively small number of cells. The approach is sophisticated enough to offer flexibility while maintaining computational efficiency, and usually it does not require advanced parallel processing. STELLA's use of visual elements to represent data and the ways it characterizes system dynamics are all closely related to pictorial simulation models (Câmara et al. 1991) and cellular automata models (Toffoli and Margolus 1987). STELLA can even run object-oriented models, although not nearly as well.

We have also used similar approaches to model large-scale, complex industrial systems, tracing flows of materials, energy, and pollutants. These models are based on engineering information and time series analysis (see, for example, Ruth et al. 2004). It is a goal to ultimately combine the dynamic engineering representations of industrial processes with accounting models to provide tools for the economic optimization of production processes.

In this book, you will learn how to build your own models to investigate potentially complex economic processes, such as trade in a barter economy, the profit-maximizing choice of inputs and outputs in a production process, and the optimal harvest of trees and fish populations. We chose STELLA because it has proven a powerful tool for building and expanding models of dynamic systems. With it, you can explore real-world systems and—perhaps even more importantly— challenge your understanding and that of others about these systems.

By now, you probably feel ready to tackle the systems described in the rest of the book, but not so fast. Practice selecting, initializing, and connecting STELLA's icons until you are familiar with the basics of this software. We will use these skills to build models of various dynamic systems in the remainder of the book.

We will start, in Chap. 2, by identifying some features fundamental to dynamic models. Because these features are basic to so many dynamic processes, they will reappear throughout the book. Then, we turn our attention to the particulars of models of optimal natural resource use. In building these models, we will focus on general principles. You are encouraged to further your own investigations by building on these basic principles to pose and solve questions concerning the dynamics of complex, real-world phenomena.

1.1.9 Basic Fish Model Equations

Here, as in later chapters of this book, we show you the model equations in STELLA. You can use them to check your model against the specifications we have chosen, and readily make adjustments as needed.

```
FISH(t) = FISH(t - dt) + (REPRODUCTION - EXTRACTION) * dt
INIT FISH = 200
INFLOWS:
REPRODUCTION = REPRODUCTION_RATE*FISH
OUTFLOWS:
EXTRACTION = EXTRACTION_RATE*FISH
EXTRACTION_RATE = .03
PRICE = 5
REPRODUCTION_RATE = GRAPH(FISH)
(2.00, 0.2), (51.8, 0.194), (102, 0.188), (151, 0.177), (201, 0.162), (251, 0.134),
(301, 0.106), (351, 0.065), (400, 0.034), (450, 0.014), (500, 0.00)
REVENUES = PRICE*EXTRACTION
```

References

Câmara AS, Ferreira FC, Fialho JE, Nobre E (1991) Pictorial simulation applied to water quality modeling. Water Sci Technol 24:275–281
Casti JL (1989) Alternate realities: mathematical models of nature and man. Wiley, New York

Meadows DH, Robinson JM (1985) The electronic oracle: computer models and social decisions. Wiley, New York

Pagels H (1988) Dreams of reason. Simon and Schuster, New York

Ruth M, Pieper F (1994) Modeling spatial dynamics of sea level rise in a coastal area. Syst Dyn Rev 10(4):375–389

Ruth M, Davidsdottir B, Amato A (2004) Climate change policies and capital vintage effects: the cases of US pulp and paper, iron and steel and ethylene. J Environ Manage 7(3):221–233

Smil V (1993) Global ecology, environmental change and social flexibility. Routledge, London

Toffoli T, Margolus N (1987) Cellular automata: a new environment for modeling. MIT Press, Cambridge

Westervelt J, Hannon B (1993) A large-scale, dynamic spatial model of the sage grouse in a desert steppe ecosystem. Mimeo, Department of Geography, University of Illinois, Urbana

Chapter 2
Disaggregation of Stocks

Logic is the muse of thought. When I violate it I am erratic; if
I hate it, I am licentious or dissolute; if I love it, I am free—
the highest blessing the austere muse can give.

C. J. Keyser, *Mathematical Philosophy*, 1922, p. 136

2.1 Promotion Within the Firm

In the previous chapter, we modeled the dynamics of a fish population living in a pond. We implicitly assumed that all fish are born at a reproductive age and leave the pond irrespective of their age. If you were operating a fishery at your pond, you may want to selectively remove fish of different age classes to maximize your revenues. This may help you to capitalize on differentials in the reproductive rates of younger and older fish and the fact that rates of reproduction are higher at lower densities. A model of the different age cohorts in your pond would require that you disaggregate the stock FISH into the different age classes and introduce age-specific reproduction and extraction rates.

Let us take up this idea of disaggregating stocks into subgroups of individuals here to model the dynamics of a company that consists of assistants, directors, and executives. Each of these three groups makes up part of your "company stock." You are put in charge of personnel in that company and are given the task of changing the distribution of people among the three positions in the company hierarchy from the current distribution to a desired one. Currently there are 800 assistants, 100 directors, and 10 executives. People get hired into the company at the assistant level, some of the assistants get promoted to become directors, and some of the directors ascent to become executives. People retire from the company after they

---

A save-disabled version of STELLA® and the computer models of this book are available at www.iseesystems.com/modelingeconomicsystems.

M. Ruth and B. Hannon, *Modeling Dynamic Economic Systems*,
Modeling Dynamic Systems, DOI 10.1007/978-1-4614-2209-9_2,
© Springer Science+Business Media, LLC 2012

reached the executive position. The desired company hierarchy is one with 900 assistants, 90 directors and 9 executives.

To achieve the desired company hierarchy, you need to hire and promote people at the appropriate rates. Assume that it is easier for you to hire new assistants than promote assistants to directors or directors to executives. Assume also that the rate of retirement is beyond your influence. The hiring rate is 100% of the difference between the actual and desired number of assistants. The rates at which you can promote assistants to directors and directors to executives are 80 and 60%, respectively. The retirement rate is 5%. Can you achieve the desired distribution for the company hierarchy within 10 years? Let us set up the STELLA model to answer this question.

Define three state variables as stocks, one each for ASSISTANTS, DIRECTORS, and EXECUTIVES. Model an inflow of HIRES into the stock of ASSISTANTS and outflows of ASSISTANTS into the stock of DIRECTORS and from DIRECTORS to EXECUTIVES. PROMOTION of ASSISTANTS to DIRECTORS is a function of the desired and actual stock and of the rate at which you can hire people into the company. The DESIRED ASSISTANTS and HIRING RATE are exogenously given parameters that you should specify as transforming variables—they will transform the flow of hires. This flow is

HIRE = HIRING RATE * (DESIRED ASSISTANTS − ASSISTANTS).   (2.1)

The larger is the difference between the number of desired and actual assistants, the more new assistants are hired. Similarly, we can define

$$PROMOTION = PROMOTION\ RATE\ * (DESIRED\ DIRECTORS \\ - DIRECTORS), \qquad (2.2)$$

$$ASCENSION = ASCENSION\ RATE\ * (DESIRED\ EXECUTIVES \\ - EXECUTIVES), \qquad (2.3)$$

$$RETIRMENT = RETIREMENT\ RATE\ * EXECUTIVES. \qquad (2.4)$$

The STELLA model is shown in Fig. 2.1.

Make an educated guess on the dynamics of the company hierarchy and then run the model over 20 years to see whether you guessed correctly and whether you can achieve the desired goals by year 10. The results are shown in Fig. 2.2.

The goal of having 900 assistants is soon achieved, but there are more directors and fewer executives than desired. Decrease the rate for promotion of assistants to directors and see whether this should help you reach the goal. What are the effects on the number of executives? Can you adjust the ascension rate to reach the desired number of executives? Where is the bottleneck in your company hierarchy, and how can you solve the problem of achieving the desired structure within 10 years? Set up the model to show, over time, the size of the male and female work force. To do this, assume different hiring, promotion and ascension, and retirement rates for the two groups of employees. Then, investigate the time it would take to have an

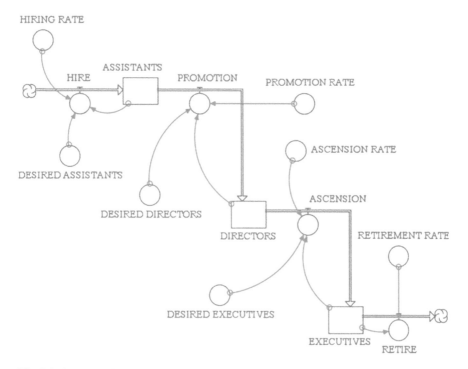

Fig. 2.1 Company hierarchy model

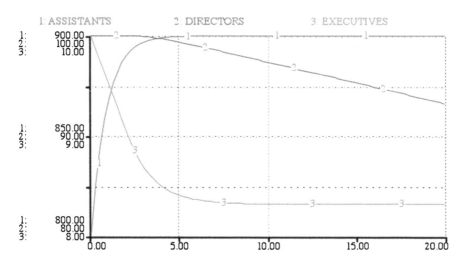

Fig. 2.2 Company hierarchy

equal share of each gender in the company. What are the implications of your findings for your personnel planning?

In this model, we have expanded on the logic of the previous chapter by tracing, for exogenously given parameters, the effects of inflows to and outflows from stocks on a system's dynamics. We have then noted the influence of the parameters on the system's state variables over time. We have disaggregated a single stock of employees into subgroups based on their positions in the company hierarchy, because treating the entire work force of the company as a single stock would not have enabled us to answer the question of optimal hiring strategies. The resolution at which we model the system depends on the question we wish to answer. The following chapter will illustrate this point in more detail. There we proceed in the opposite direction. Rather than disaggregating a system of interest into ever-smaller parts, we increasingly add to the model parts of the system's surroundings.

2.2   Disaggregated Stocks Model Equations

ASSISTANTS(t) = ASSISTANTS(t - dt) + (HIRE - PROMOTION) * dt
INIT ASSISTANTS = 800
INFLOWS:
HIRE = (DESIRED_ASSISTANTS-ASSISTANTS)*HIRING_RATE
OUTFLOWS:
PROMOTION = (DESIRED_DIRECTORS-DIRECTORS)*PROMOTION_RATE

DIRECTORS(t) = DIRECTORS(t - dt) + (PROMOTION - ASCENSION) * dt
INIT DIRECTORS = 100
INFLOWS:
PROMOTION = (DESIRED_DIRECTORS-DIRECTORS)*PROMOTION_RATE
OUTFLOWS:
ASCENSION = (DESIRED_EXECUTIVES-EXECUTIVES)*ASCENSION_RATE

EXECUTIVES(t) = EXECUTIVES(t - dt) + (ASCENSION - RETIRE) * dt
INIT EXECUTIVES = 10
INFLOWS:
ASCENSION = (DESIRED_EXECUTIVES-EXECUTIVES)*ASCENSION_RATE
OUTFLOWS:
RETIRE = EXECUTIVES*RETIREMENT_RATE

ASCENSION_RATE = .6
DESIRED_ASSISTANTS = 900
DESIRED_DIRECTORS = 90
DESIRED_EXECUTIVES = 9
HIRING_RATE = 1
PROMOTION_RATE = .8
RETIREMENT_RATE = .05

Part II
Methods for Dynamic Modeling

Chapter 3
System Boundaries in Space and Time

I know that history at all times draws strangest consequence from remotest cause.

T. S. Eliot, *Murder in the Cathedral*, 1935, Part I.

3.1 Introduction

Much of the debate about the adequacy of particular models for describing real-world phenomena stems from the fact that we all view the world differently and thus place different importance on the processes that we observe. Some of these differences then manifest themselves in the selection of particular aspects of thereal world for our models. Through this selection process we effectively define the boundaries of the system we wish to investigate with our model. System boundaries delineate what we consider important from what we deem unimportant. We model the important part.

We wish to distinguish two types of system boundaries that are related to each other. Boundaries in space delineate the system under investigation from its surroundings. We are quite familiar with the use of such boundaries. They are drawn on maps to delineate different countries, or oceans from coastlines. But recognize, many of these boundaries are disputed or change as the resolution of our map changes. Similarly, we can delineate companies or households as individual entities—or systems—based on the type of economic activities performed by them and by property rights, location, or other criteria. These boundaries, too, can be subject to considerable debate, as it is well exemplified in the case of multinational corporate systems—conglomerates that have offices in different countries are "owned" by companies or shareholders in different countries, and perform a multitude of tasks.

A save-disabled version of STELLA® and the computer models of this book are available at www.iseesystems.com/modelingeconomicsystems.

M. Ruth and B. Hannon, *Modeling Dynamic Economic Systems*,
Modeling Dynamic Systems, DOI 10.1007/978-1-4614-2209-9_3,
© Springer Science+Business Media, LLC 2012

Even more difficult to deal with than boundaries in space are boundaries in time. The impacts that a system as on its surroundings may not be immediate and possibly rather complex, following a chain of interrelated responses in its environment that then may feed back to the system itself. Which of those impacts should be captured in our model? Surely, those impacts that are comparably small and that do not occur for a long time may be safely regarded as irrelevant to understanding the system's performance. Our model will then operate under a set of ceteris paribus assumptions—meaning that we postulate that "everything that is not part of the model remains the same" as our system moves from one state to another. But, what do we mean when we say that impacts are "comparably small" and do not occur "for a long time"? Which of the ceteris paribus assumptions can be justified and which cannot? An experimental approach to modeling can help us make this decision by laying open the relevance of our assumptions.

In the following sections of this chapter, we subsequently extend the boundaries of our model. We start with an individual firm and ultimately capture activities of other firms and part of the resource base. Of course, a large number of potentially important real-world features will not be included in this model. We abstract away from them. However, as we extend our model, we find that our answers to the question that the model is supposed to answer will change.

3.2 Energy Cost of Production at the Level of the Firm

Let us consider a single firm that produces a single good as its output. The quantity Q of the good is generated by using materials and energy. The firm knows—or thinks it knows—all future sales of its good and produces just as much as can be sold for a given price. The expected sales and production may be thought as following a product life cycle pattern with low sales early on as the good is introduced in the market, increasing sales as the market develops, and declining sales the longer the good has been available. Declines in sales by this firm may be due to increased competition from other firms that enter the market. We do not model market access explicitly here, but will take up this issue in some detail in Chap. 11.

Many goods seem to follow a life cycle pattern. Black-and-white television sets, pocket calculators, and turntables for stereo equipment are examples of goods that have been around for a while and whose sales and sales potential declined quite a while ago. Mobile phones may have already approached their peak in sales while electronic book readers are still at relatively early stages of their product life cycle. For our model, we capture the sales over time with the simplified, hypothetical relationship shown in Fig. 3.1.

Fig. 3.1 Sales profile over a
product lifecycle

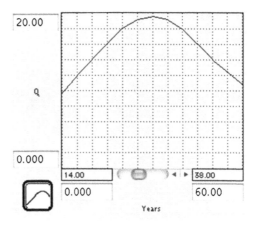

The entire data for the above figure is shown here:

0	2.10
2	2.30
4	2.90
6	3.70
8	4.70
10	6.00
12	7.70
14	9.60
16	11.90
18	14.00
20	16.10
22	18.10
24	19.20
26	19.60
28	19.20
30	18.00
32	16.10
34	14.00
36	12.50
38	10.90
40	9.50
42	8.20
44	7.10
46	6.30
48	5.40
50	4.50
52	3.70
54	2.80
56	2.30
58	1.90
60	1.90

Fig. 3.2 Declining energy
requirements per unit output
(*EQ\Q*) as cumulative
production (*Z*) increases

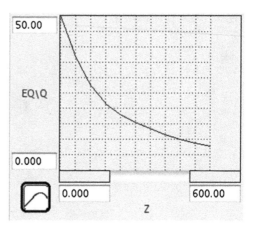

When you open the STELLA model, which you can download (for details
see Chap. 1), look at the transforming variable *Q* that captures this relationship in
a graph. Press the Option (Macintosh) or Control (Windows) key to view the entire
graph.

Bell-shaped trends, as shown in the life cycle pattern here, often result from an
interplay of at least two counteracting forces whose relative strength changes over
time. In our case, one of these forces is the increase in the size of the market
or development of technologies to produce the goods at ever-lower cost. The other
force is the competitors whose entry into the market drives down the sales of the
firm. But recognize, the result of these counteracting forces does not necessarily
have to be a bell-shaped curve, as we assumed here. With an increased number of
feedback processes and time lags in the response of system components, many
variations and reversals of trends can occur, as we will see in later chapters.
Through our choice of system boundaries, however, we do not deal with such
feedback processes and time lags.

Let us assume that the firm anticipates that, with increased experience in
producing the good, it will be able to increase its efficiency of using energy in the
production process. Experience may be expressed through cumulative production,
Z. Energy efficiency is measured as energy use per unit output, *EQ\Q*. The relationship is given in Fig. 3.2.

The question that we wish to answer with our model is the following: What is the
total and per unit energy consumed over time to produce the output of the firm? To
provide an answer, the model must keep track of cumulative production *Z*, and must
calculate energy use, *EQ*, in the production of *Q*. Energy use is just the product of
per unit energy consumption and output:

$$EQ = EQ\backslash Q * Q. \tag{3.1}$$

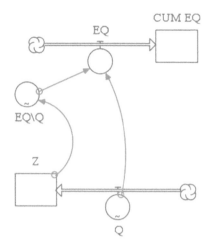

Fig. 3.3 Simple energy use model

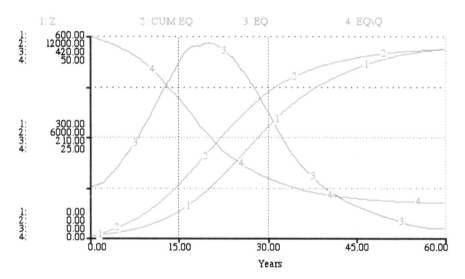

Fig. 3.4 Results of the simple energy use model

Our model is shown in Fig. 3.3.

Set the initial values of cumulative production and cumulative energy use to 1. Choose a DT = 0.25 and run the model over 60 years of simulation time. Figure 3.4 shows what you should get. Cumulative production rises at an increasing rate early on but levels off as output declines. The increase in cumulative production is accompanied by a decline in energy costs of production $EQ\backslash Q$. The combined effect of declining energy cost and increased rates of production temporarily results in increased energy use EQ. Energy use ultimately declines as production levels off.

The firm that we modeled here can pride itself with the fact that it achieved more than an 80% decline in energy use per unit output over the product life cycle. Change the initial values of Z and note the effects on the results. In reality it is often difficult to know the entire cumulative past production of a firm, unless you have access to all its production records. By truncating that past and (arbitrarily) assuming an initial value for Z, we made an assumption on the system boundaries in time. You should find that this assumption is not very crucial for the model results, provided that we indeed deal with a product early in its product life cycle.

3.3 Firm-Level Energy Cost Model Equations

CUM_EQ(t) = CUM_EQ(t - dt) + (EQ) * dt
INIT CUM_EQ = 1
INFLOWS:
EQ = EQ\Q*Q {MJ}

Z(t) = Z(t - dt) + (Q) * dt
INIT Z = 1
INFLOWS:
Q = GRAPH(TIME)
(0.00, 2.10), (2.00, 2.30), (4.00, 2.90), (6.00, 3.70), (8.00, 4.70), (10.0, 6.00),
(12.0, 7.70), (14.0, 9.60), (16.0, 11.9), (18.0, 14.0), (20.0, 16.1), (22.0, 18.1),
(24.0, 19.2), (26.0, 19.6), (28.0, 19.2), (30.0, 18.0), (32.0, 16.1), (34.0, 14.0),
(36.0, 12.5), (38.0, 10.9), (40.0, 9.40), (42.0, 8.20), (44.0, 7.10), (46.0, 6.30),
(48.0, 5.40), (50.0, 4.50), (52.0, 3.70), (54.0, 2.80), (56.0, 2.30), (58.0, 1.90),
(60.0, 1.90)

EQ\Q = GRAPH(Z {MJ/Ton})
(0.00, 50.0), (60.0, 36.8), (120, 27.5), (180, 21.5), (240, 18.0), (300, 15.5), (360,
13.5), (420, 11.5), (480, 10.0), (540, 8.75), (600, 7.75)

3.4 Extending the System Boundaries

In the previous section, we conceptually drew the system boundaries to capture a single firm's output and the corresponding energy use. We assumed that the firm was able to increase energy efficiency as it gained more experience. Let us introduce a similar learning effect for the use of materials. Material use per unit output is denoted $X\backslash Q$ and represented as a graphical function of cumulative production (Fig. 3.5). The more of the good has been produced in the past, the more experience has been gained and thus the lower are the material input requirements per unit output.

Assume this material must be extracted from the ground, such as copper or iron ore. Increased material use, thus, corresponds to increased extraction rates.

Typically, companies extract minerals from their best reserves first before they move on to those that are of a lower quality. The quality of the mineral reserves is reflected by its ore grade and its expected extraction cost. The lower the grade, the

Fig. 3.5 Declining material use per unit output ($X\backslash Q$) with increasing cumulative production (Z)

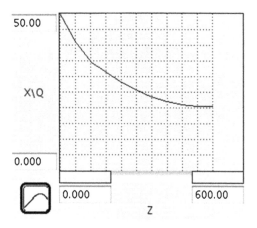

lower the quality—more impurities and waste materials have to be removed to yield a ton of metal. Historically, ore grades of many metals dropped significantly. For example, at the beginning of this century copper ore grades in the United States were approximately 2% pure metal content. Today, they are less than 0.5%. Similarly, iron ore grades dropped from close to 40% in 1950 to less than 20% in 1995.

Let us model changes in ore grade G as a *declining* function of cumulative material use, CUM X. The functional form is

$$G = \text{ALPHA1} * \text{CUM } X^{\wedge}\text{ALPHA2}. \tag{3.2}$$

We arbitrarily set the constants ALPHA1 $= 0.7$ and ALPHA2 $= -0.2$. Note that we could have portrayed G as a graph in one of STELLA's converters and used multipliers on the graphical input or output to simulate the effect of ALPHA1 and ALPHA2. Other ways of dealing with such depletion effects are presented in detail in Part IV of the book.

Of course, making input materials available for the production of Q itself requires energy. Energy use in material extraction per unit of the material extracted typically declines as ore grades drop. We model this relationship between ore grade G and energy use per unit X, $EX\backslash X$, with the following function:

$$EX\backslash X = \text{BETA1} * G^{\wedge}\text{BETA2}, \tag{3.3}$$

with BETA1 $= 0.8$ and BETA2 $= -0.6$ as constants. As ore grade changes over time due to continued material use for the production of Q, energy use EX in the extraction of X increases as G drops. These relationships are captured in the part of the STELLA model illustrated in Fig. 3.6.

The total energy use E associated with the production of Q—*directly* in the production of Q itself and *indirectly* for the extraction of X—is

$$E = EQ + EX. \tag{3.4}$$

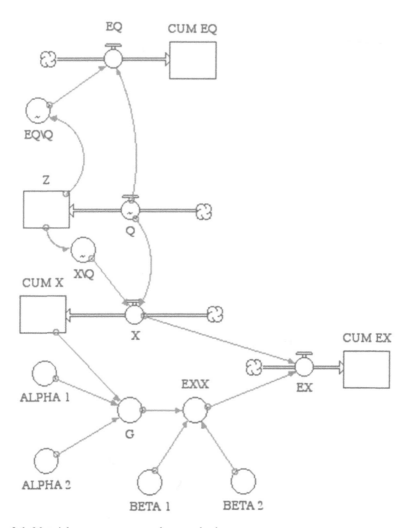

Fig. 3.6 Material use, energy use, and ore grade changes

Dividing E by the total amount of Q yields the total of direct and indirect energy required in the production of Q, $E\backslash Q$.

With each additional feature, the diagram of the model increases in size. In larger models that have highly interdependent components we need to make a large number of connections. With an increasing number of connections, or information arrows, the readability of the model can be seriously reduced. Use "Ghosts" of icons to avoid crossing arrows and increase the transparency of the model structure. You can create a ghost, for example, of a converter by first clicking on the ghost icon shown in Fig. 3.7.

Once you clicked on the ghost icon, move it to the variable that you want to duplicate. Click on the symbol you want to duplicate. The ghost icon then changes

Fig. 3.7 Ghost

Fig. 3.8 Energy use

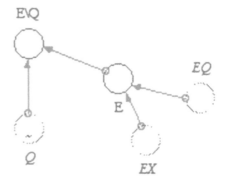

its appearance into that of the symbol you clicked on. You can now place this duplicate of the original anywhere in the diagram and connect it with information arrows to the relevant parts of the model. The ghost you created is the same icon as its original, but printed slightly lighter in color than the original.

In our model, we created a duplicate, or "ghost," of *EQ* and *EX* to easily connect them to calculate *E*. Similarly, we used a ghost of *Q* to calculate *E\Q*. The ghost always assumes the value of the original. Be aware that all changes to the variable must be made to the original, not the ghost. Thus, information arrows can originate from the ghosts to connect them to other parts of the model, but it is not possible that ghosts receive information arrows (Fig. 3.8).

Run the model for initial values of *Z*, CUM *EQ*, and CUM *EX* equal to 1 and CUM *X* equal to 10. The cumulative values of output, material use, and energy use are shown in Fig. 3.9. All curves show initially fast increases due to the rapid expansion in *Q*.

Figure 3.10 shows the life cycle pattern of production *Q*, material use *X,* and material use per unit output. Because production of *Q* ultimately levels out, so does the use of materials. The decline in ore grade G with increased cumulative extraction and use of materials *X* is shown in Fig. 3.11. Ore grades initially drop fast. As less material is extracted—because of declining *Q* and *X\Q*—ore grades decline at a lower rate. Figure 3.12 shows changes in energy use per unit output. Energy use per unit material extracted increases steadily because ore grades decline. By contrast, energy use per unit output *Q* declines. The joint effect of these two opposite trends in energy use is a bell-shaped path for the combined direct and indirect energy consumption per unit output.

Let us now return to the question we posed in the previous section: What is the total and per unit energy consumed over time to produce the output of the firm? Compare the graphs above with the results of the model in the previous section, where we were concerned with only the firm producing *Q*. Now that

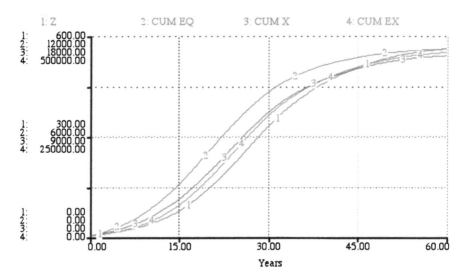

Fig. 3.9 Cumulative output, material use, and energy use

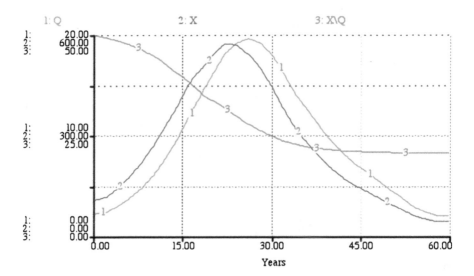

Fig. 3.10 Output and material use

we included the effects of material use by that firm on energy use elsewhere in the system, we must be more careful when judging energy efficiency improvements by the firm. Could the temporary rise in the combined direct and indirect energy use per unit output be reduced if the firm had improved material use efficiencies $X\backslash Q$ a bit more, even if this improvement came at the expense of an increase in $EQ\backslash Q$?

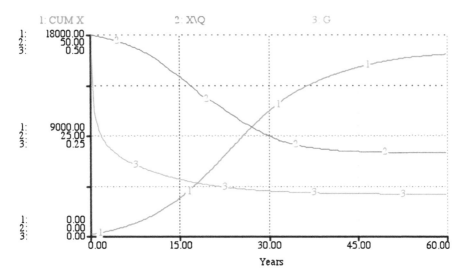

Fig. 3.11   Changes in material use

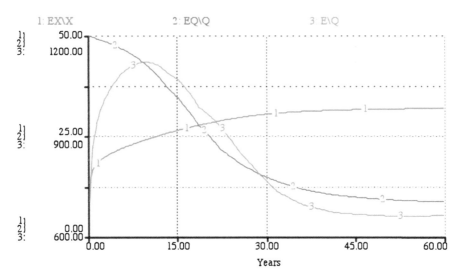

Fig. 3.12   Changes in material and energy uses per unit output

Of course, our answer is different here from the case in the previous section because we included the effects that material use has on ore grades and through ore grades on energy use in the "extractive sector." A number of different answers to our question can be given, depending on the choice of the parameters that we used. However, the answer depends not only on the choice of parameters but also, to a very significant extent, on our choice of the system boundaries, as the last of the preceding graphs illustrates.

Rerun the model with alternative initial values of CUM X and alternative parameters for the functional forms that capture the relationship between ore grade and cumulative material use, (3.1), and between ore grade and energy cost of extraction, (3.2). Modify (3.1) and (3.2) to include effects of the rates of material use X on ore grades and energy that is required to produce a unit of output. What should the direction of this effect be?

3.5   Extended System Boundary Model Equations

```
CUM_EQ(t) = CUM_EQ(t - dt) + (EQ) * dt
INIT CUM_EQ = 1
INFLOWS:
EQ = EQ\Q*Q  {MJ}

CUM_EX(t) = CUM_EX(t - dt) + (EX) * dt
INIT CUM_EX = 10
INFLOWS:
EX = EX\X*X

CUM_X(t) = CUM_X(t - dt) + (X) * dt
INIT CUM_X = 10
INFLOWS:
X = X\Q*Q

Z(t) = Z(t - dt) + (Q) * dt
INIT Z = 1
INFLOWS:
Q = GRAPH(TIME)
(0.00, 2.10), (2.00, 2.30), (4.00, 2.90), (6.00, 3.70), (8.00, 4.70), (10.0, 6.00),
(12.0, 7.70), (14.0, 9.60), (16.0, 11.9), (18.0, 14.0), (20.0, 16.1), (22.0, 18.1),
(24.0, 19.2), (26.0, 19.6), (28.0, 19.2), (30.0, 18.0), (32.0, 16.1), (34.0, 14.0),
(36.0, 12.5), (38.0, 10.9), (40.0, 9.40), (42.0, 8.20), (44.0, 7.10), (46.0, 6.30),
(48.0, 5.40), (50.0, 4.50), (52.0, 3.70), (54.0, 2.80), (56.0, 2.30), (58.0, 1.90),
(60.0, 1.90)

ALPHA_1 = .7
ALPHA_2 = -.2
BETA_1 = 8
BETA_2 = -.6
E = EQ+EX  {MJ}
EX\X = BETA_1*G^BETA_2  {MJ per Ton}
E\Q = E/Q  {MJ per Ton}
G = ALPHA_1*CUM_X^ALPHA_2  {Tons of X per Ton of Ore Extracted}
EQ\Q = GRAPH(Z  {MJ/Ton})
(0.00, 50.0), (60.0, 36.8), (120, 27.5), (180, 21.5), (240, 18.0), (300, 15.5), (360,
13.5), (420, 11.5), (480, 10.0), (540, 8.75), (600, 7.75)
X\Q = GRAPH(Z)
(0.00, 50.0), (60.0, 41.0), (120, 34.8), (180, 31.5), (240, 28.2), (300, 25.8), (360,
23.5), (420, 22.0), (480, 21.0), (540, 20.5), (600, 20.5)
```

3.6 Sensitivity Analysis

There are many directions in which the model of the previous section can be refined and extended to make it more realistic. We could specifically model the use of other inputs in the production process such as labor and capital, and deal explicitly with the possibilities of trade-offs in the use of one of these inputs for others. For example, to some extent capital goods may be used to substitute for labor or material inputs. Obviously, capital goods themselves need to be produced, using materials and energy. An interesting question that arises from such a model is whether increased capital use actually helps decrease total material and energy requirements in the system. Return to this question after you worked through this section.

Real production processes frequently show possibilities for substitution—at least within certain realms. We will focus on the realm of best input substitution possibilities in Chap. 8.

Another extension to the model is to include the energy cost of extracting energy. So far, we were concerned with only the energy cost of extracting materials. But, of course, energy must be extracted in the first place to be able to extract and process the materials, and energy extraction itself requires energy. In this section, we extend our model in this direction and then introduce a systematic procedure to test for the sensitivity of the model results to our choice of parameters.

The combined effects of substitution and resource depletion on the energy requirements per unit of energy and material extraction can be significant. For example, use of wood for heating, cooking, and construction in England led to a significant depletion of forest resources in the seventeenth century. Increases in economic activity and in population during the industrial and agricultural revolutions required increased use of energy. Charcoal, produced from wood, was increasingly used to supply this energy, which exacerbated the problems in wood supply. As pressures on remaining forest resources increased further, coal— an abundant energy source of high heat content—was used as a substitute. Its chemical and physical characteristics made it a preferred input in iron smelting and refining, processes that became increasingly important for the industrial revolution. However, as more and more coal was extracted from England's coal mines, more energy had to be used to extract coal, and to pump water that infiltrated the mine shafts and tunnels. At increased depths, human and animal labor was insufficient in handling the large amounts of water that had to be pumped from the mines. The invention of the steam engine provided a technological breakthrough that helped alleviate some of the problems in mining. However, steam engines required coal to operate. As a result even more energy that was extracted from the mines was diverted to the extraction process itself. The steam engine not only revolutionized mining of coal but also of iron ore. More materials were transported from mines to smelters, refineries, and population centers. The steam engine, through its applications on railroads, facilitated the transportation of those materials. But again, more energy was required to move goods and services, to produce iron and steel for railroads and other uses, and to extract the coal used to extract iron ore and coal itself.

Let us model in a simplified way the energy requirements by the energy extracting sector. Assume that the energy required per unit energy extracted $EE\backslash E$ depends on cumulative energy extraction, CUM EE. The relationship is similar to the one used above to define changes in ore grade of materials with cumulative material extraction. The functional form used in the model is

$$EE\backslash E = \text{GAMMA1} \,^* \text{ CUM } EE^{\wedge}\text{GAMMA2}, \tag{3.5}$$

where GAMMA1 $= 2$ and GAMMA2 $= 0.1$. Unlike the case of materials, where increased cumulative material extraction drove down ore grades, increased cumulative energy extraction *increases* the energy used in extracting the next unit of energy.

Let us assume that the demand for energy in the model is composed of the demand for energy only to extract materials, to process them in the production of Q units of the final product, and to extract energy itself. For simplicity, we assume here that own energy use by the energy sector is directly proportional to the energy use in material extraction and production of Q. The proportionality factor A is 10%. Therefore, total energy extraction is

$$EE = (1 + A) \,^* EE\backslash E \,^* (EQ + EX). \tag{3.6}$$

Indirect energy use to produce Q now includes the energy to extract energy and the energy to extract materials. The direct energy for the production of Q is EQ. To calculate the total direct and indirect energy use in the system per unit output we now need to divide total demand for energy EE by Q. Our extended model now consists of the parts shown in Fig. 3.13.

Energy use per unit output of the materials extraction sector, the energy extraction sector, and the firm producing the final product are shown in Fig. 3.14. We also calculated the total direct and indirect energy per unit output of the final product. Compare the results with the previous section, and you will find that the inclusion of energy cost in the energy-extracting sector significantly increased the total direct and indirect energy use per unit output, $E\backslash Q$.

Throughout this chapter, we have stressed the dependency of the model results on system boundary definitions and parameter choice. We have explored the role of system boundaries by including in subsequent model development stages parts of the system that we assumed had significant impacts on our results. Let us now turn to the sensitivity of model results to the parameter values used in the model.

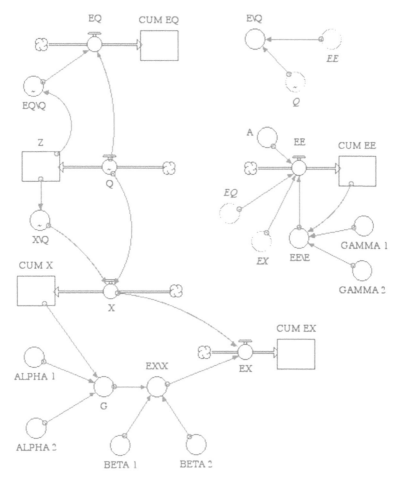

Fig. 3.13 Extended model

The last model extension showed that the inclusion of the energy use in energy extraction determines to a significant extent total direct and indirect energy use per unit of the finished product. Here we assumed that the energy-extracting sector itself uses 10% of the demand by other sectors in the model. How sensitive are the model results to this assumption?

Let us vary the parameter A, which captures the percentage of own energy use in the energy extracting sector for consecutive runs, and plot all the resulting curves of $E\backslash Q$ together in the same graph. To do this, we make use of sensitivity runs, a convenient method provided by the STELLA software.

Go to the RUN pull-down menu, select "Sensi Specs" (Sensitivity Specifications), and choose the parameter A—by clicking on it and selecting it—as the one

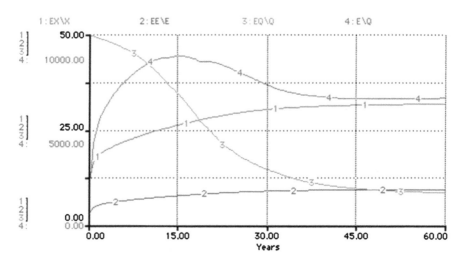

Fig. 3.14 Results of the extended model

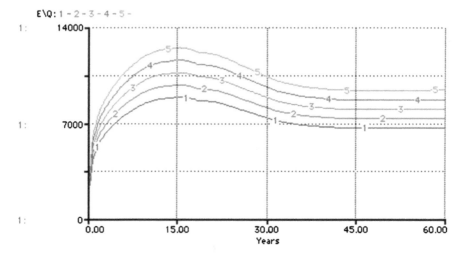

Fig. 3.15 Sensitivity analysis

for which we want to perform a sensitivity analysis. Type in the dialogue box "# of Runs" 5 to generate five sensitivity runs. Then provide start and end values for *A*. *A* assumes the start value for the first sensitivity run and the end value for the last sensitivity run. If you chose "Incremental" as the variation type, STELLA will calculate the other values from the start and end values that you specified such that there are equal incremental changes in *A* from run to run. Plot the five resulting

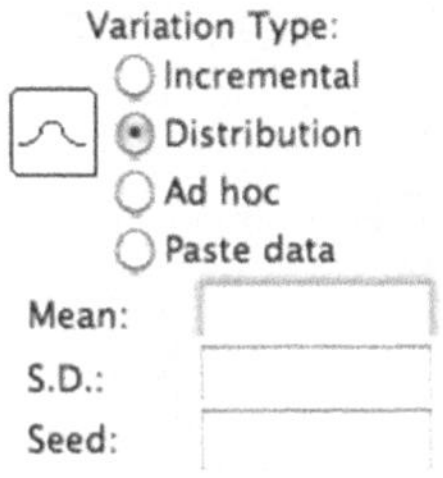

Fig. 3.16 Specification of
sensitivity analysis with
normally distributed
parameter values

Variation Type:
○ Incremental
◉ Distribution
○ Ad hoc
○ Paste data

Mean: [   ]
S.D.: [   ]
Seed: [   ]

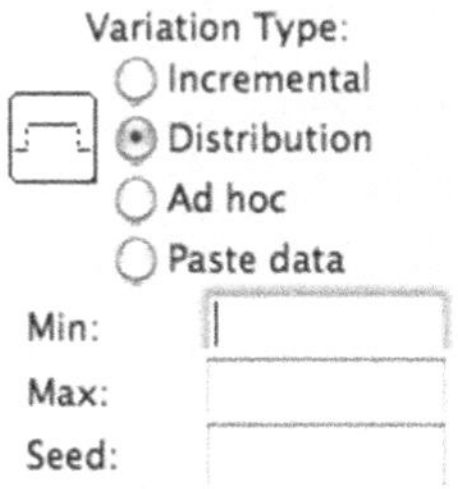

Fig. 3.17 Specification of
sensitivity analysis with not
normally distributed
parameter values

Variation Type:
○ Incremental
◉ Distribution
○ Ad hoc
○ Paste data

Min: [   ]
Max: [   ]
Seed: [   ]

curves for $E\backslash Q$ in the same graph by choosing the "Graph" option in STELLA's
SENSITIVITY SPECS menu. Run the model with the S-RUN command and
observe the resulting graph (Fig. 3.15).

Obviously our assumption on own energy use by the energy extraction sector
does not drastically influence the results of the model. Choose other parameters in
the model and assess the sensitivity of the results to those parameters. If you wish
to let them vary from run to run along a normal distribution with a known mean and
standard deviation, choose the "Distribution" option instead of "Incremental"
(Fig. 3.16). When you specify "Seed" as a positive integer, you ensure the ability
to replicate a particular random number sequence in subsequent sensitivity runs.
Specify 0 as the seed, and a "random" seed will be selected. If you do not wish to
utilize the normal distribution of the random numbers used in the sensitivity
analysis, click on the bell-curve button (Fig. 3.16).

This button will change the curve's appearance (Fig. 3.17), and you now need to
specify a minimum, maximum, and a seed for your sensitivity analysis.

A final choice for the specification of sensitivity runs is not to use STELLA's
prescribed changes in parameters in incremental intervals or along distributions.
You can specify ad hoc values for each of the consecutive runs. These options
should provide you with sufficient flexibility in the assessment of the sensitivity
of the model to the numerical values you choice to initialize it. Make frequent
use of sensitivity runs. They will help you better understand your model and its
reliability.

3.7   Sensitivity Analysis Model Equations

CUM_EE(t) = CUM_EE(t - dt) + (EE) * dt
INIT CUM_EE = 100
INFLOWS:
EE = (1+A)*EE\E*(EQ+EX) {MJ per Year}

CUM_EQ(t) = CUM_EQ(t - dt) + (EQ) * dt
INIT CUM_EQ = 1
INFLOWS:
EQ = EQ\Q*Q {MJ}

CUM_EX(t) = CUM_EX(t - dt) + (EX) * dt
INIT CUM_EX = 10
INFLOWS:
EX = EX\X*X

CUM_X(t) = CUM_X(t - dt) + (X) * dt
INIT CUM_X = 10
INFLOWS:
X = X\Q*Q

Z(t) = Z(t - dt) + (Q) * dt
INIT Z = 1
INFLOWS:
Q = GRAPH(TIME)
(0.00, 2.10), (2.00, 2.30), (4.00, 2.90), (6.00, 3.70), (8.00, 4.70), (10.0, 6.00),
(12.0, 7.70), (14.0, 9.60), (16.0, 11.9), (18.0, 14.0), (20.0, 16.1), (22.0, 18.1),
(24.0, 19.2), (26.0, 19.6), (28.0, 19.2), (30.0, 18.0), (32.0, 16.1), (34.0, 14.0),
(36.0, 12.5), (38.0, 10.9), (40.0, 9.40), (42.0, 8.20), (44.0, 7.10), (46.0, 6.30),
(48.0, 5.40), (50.0, 4.50), (52.0, 3.70), (54.0, 2.80), (56.0, 2.30), (58.0, 1.90),
(60.0, 1.90)

A = .1 {MJ per MJ}
ALPHA_1 = .7
ALPHA_2 = -.2
BETA_1 = 8
BETA_2 = -.6
EE\E = GAMMA_1*CUM_EE^GAMMA_2
EX\X = BETA_1*G^BETA_2 {MJ per Ton}
E\Q = EE/Q
G = ALPHA_1*CUM_X^ALPHA_2 {Tons of X per Ton of Ore Extracted}
GAMMA_1 = 2
GAMMA_2 = .1
EQ\Q = GRAPH(Z {MJ/Ton})
(0.00, 50.0), (60.0, 36.8), (120, 27.5), (180, 21.5), (240, 18.0), (300, 15.5), (360,
13.5), (420, 11.5), (480, 10.0), (540, 8.75), (600, 7.75)
X\Q = GRAPH(Z)
(0.00, 50.0), (60.0, 41.0), (120, 34.8), (180, 31.5), (240, 28.2), (300, 25.8), (360,
23.5), (420, 22.0), (480, 21.0), (540, 20.5), (600, 20.5)

Chapter 4
Scheduling Flows

Nothing is permanent but change.

Heraclitus 500 B.C.

4.1 Conveyors, Queues, and Ovens

In the previous chapters we used stocks to represent state variables. The corresponding STELLA symbol is the Reservoir with inflow and outflow at each DT. In reality, however, many stocks receive inflows and result in outflows only at particular periods of time, depending on a "schedule" that underlies the processes of the respective system. The scheduling of inflows into and outflows from stocks can be easily modeled by using the other versions of stocks offered in STELLA: Conveyor, Queue, and Oven. To create any of these stocks, select the STELLA symbol of a stock (Fig. 4.1), drag it into the diagram, and open it by double-clicking on it. At the top of the dialogue box you can choose the type of stock.

A Conveyor is able to take up some value or values of state variables, keep these values for a fixed length of time, and then release the values describing the state variable. Each DT is assigned a slat on the conveyer. Conveyors can be compared to conveyor belts on which some object moves before it gets off again. Whatever goes into the Conveyor first also leaves the Conveyor first. The same holds for Queues. These are lines of objects that await entry into some process. Objects first in line, that is, arriving first in the queue, will enter the process first. Lines of people waiting at a grocery store checkout are an example.

Ovens are processors of discrete batches of objects, very much like ovens used in a bakery. Ovens receive a number of objects over some period of time, retain these objects for some time, and then unload them all in an instant.

A save-disabled version of STELLA® and the computer models of this book are available at www.iseesystems.com/modelingeconomicsystems.

M. Ruth and B. Hannon, *Modeling Dynamic Economic Systems*,
Modeling Dynamic Systems, DOI 10.1007/978-1-4614-2209-9_4,
© Springer Science+Business Media, LLC 2012

Fig. 4.1 Alternative
representations of stocks

Although we are used to thinking of objects as discrete entities—such as one
car, two telephones—many models assume divisibility of those entities. For example,
infinite divisibility of units is frequently assumed in economic models. Consumers,
for example, may increase their utility by receiving a quarter of a car. Similarly, firms
may increase their profits to a maximum by selling half a telephone more.

With increased use of the calculus of variations in economic models, assumptions of infinite divisibility of units were made necessary, and as a result, discrete
processes have been frequently neglected. However, real-world production and
consumption processes are replete with examples of discrete units. Conveyors,
ueues, and Ovens are forms of stocks that are designed to model such discrete
processes, and some examples of their use follow.

This chapter presents two models of discrete processes. The first of these models
deals with a simple grocery store. This model introduces the tools for modeling
discrete processes and shows how to handle multiple independent variables in
STELLA. Time is our usual independent variable, and we use it to cause the change
in our models. In the next sections, we model systems not only with respect to their
temporal dimension but also with respect to their spatial dimension. By doing so, we
model two independent variables—space and time. The methods of modeling
discrete processes in space and time are used in the final section of this chapter to
optimize a system's performance. Here, we identify an optimal schedule for a public
transportation system to minimize travel time. The same tools and methods can be
used to optimize, for example, discrete production processes of individual firms.

4.2   Modeling Discrete Flows in Space and Time

To illustrate the use of conveyors, queues, and ovens, consider the simple case of a
store into which people enter, receive some service, then move to the cash register, and
have to wait in a checkout line before they can pay and leave. Only one person can be
served at a time, and initially one person is already at the service center being served. It
takes 5 min to be served and 1 min to get from the service center to the checkout line.

The "inflow" of people into the store can be modeled, as before, with a unidirectional
flow. Let us call it ENTER and specify it later. Upon arrival, people wait in line, the
SERVICE QUEUE, to be served at the SERVICE CENTER. Place two stocks in your
STELLA II diagram, one called SERVICE QUEUE and the other called SERVICE
CENTER. Connect both with a flow called GET SERVICE. Next, place a stock on the
diagram, call it TRANSFER, and connect it with a flow to the SERVICE CENTER.
Choose another stock for the checkout line, with inflow from TRANSFER and outflow
into a cloud to represent the people leaving the store. The final state variable,
represented by a stock, is the CASH REGISTER, capturing the people currently paying.

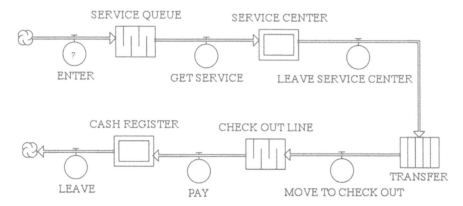

Fig. 4.2 Store model

Now that we have laid out the model, let us specify the stocks and flows. Open the SERVICE QUEUE stock and specify it as a queue with an initial value of 2, assuming two people are already waiting to be served. Recognize that once you click on OK, the question mark will disappear, not only in the stock icon but also in the GET SERVICE flow. STELLA recognizes that the outflow from an oven is determined by its cook time.

Next, open the SERVICE CENTER stock and specify it as an oven with an initial value of 1, for the person already being served, and a "Cook Time" of 5. The cook time of 5 min defines the duration of items in the oven.

Similarly, specify the TRANSFER stock as a conveyor, set its initial value to 0—assuming no people are currently on their way from the service center to the checkout line—and the transit time as 1, corresponding to the 1 min it takes to walk to the checkout line. Click on OK, and the question mark in MOVE TO CHECK-OUT will disappear.

Now specify the state variable CHECKOUT LINE as a queue, with an arbitrary initial value of 8, for the eight people waiting in line. When you click OK, the question mark in the outflow PAY will disappear. People coming out of the checkout line are assumed to pay at the CASH REGISTER before they can leave the store. The CASH REGISTER is specified as an oven with a cook time of 2 min, and we assume that currently one person is being served.

Your model should now look like the one in Fig. 4.2, and we are left to specify the rate of arrival of customers into the store.

Let us assume that one customer arrives every 4 min and that the first customer arrives in the third minute after we began to run the model. The PULSE built-in function is designed to model such a case. We specify

$$\text{ENTER} = \text{PULSE}(1, 3, 4), \tag{4.1}$$

where the first entry in the PULSE function refers to the volume of the pulse, the second to the initial occurrence of the pulse, and the third entry to the length of the recurrence interval. Additionally, we "time-stamp" the inflow into the store and the outflows from the store. This is done by checking the "Time-stamp check box" in the upper right-hand corner of the ENTER flow diagram (Fig. 4.3).

Fig. 4.3 Time-stamp
elements

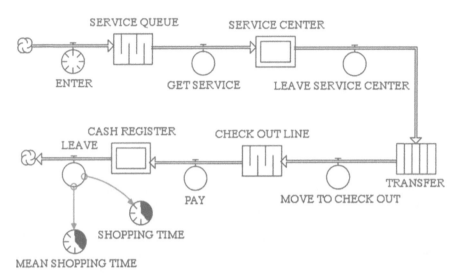

Fig. 4.4 Store model with time stamp

After you checked the box and clicked OK, the flow's appearance will change to indicate that it has been time-stamped (Fig. 4.4).

We are now able to calculate the time it takes to shop. We do so with the built-in function Cycle time. This built-in function requires for its specification, as its first input, a flow whose cycle time is calculated and, as a second input, a weight. When no weight is specified, Cycle time will return the per batch cycle time independent of volume. When the weight is set to 1, it will return the per unit volume cycle time. The latter specification is chosen in our model:

$$\text{SHOPPING TIME} = \text{ CYCLE TIME(LEAVE).} \tag{4.2}$$

Similarly, we can calculate the average cycle time per unit volume cycles as

$$\text{MEAN SHOPPING TIME} = \text{CT MEAN(LEAVE}, 1). \tag{4.3}$$

The complete model is shown in the following diagram. The results of our model, presented in the graph, show overall a stepwise decrease in the CHECK-OUT LINE, with short temporary increases, and increases in the queue prior to the SERVICE CENTER. Given an initially long CHECKOUT LINE, the time it takes an individual customer to shop is initially high, declines temporarily, and as more shoppers arrive, increases again. By the same token, the average shopping time declines temporarily and increases the longer the queue gets in front of the service center (Figs. 4.5 and 4.6).

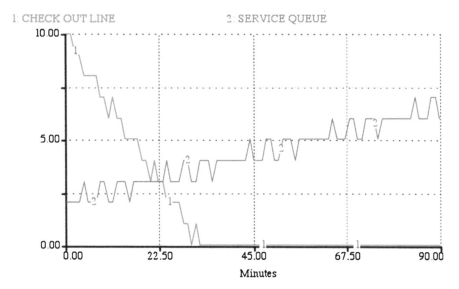

Fig. 4.5 Waiting times

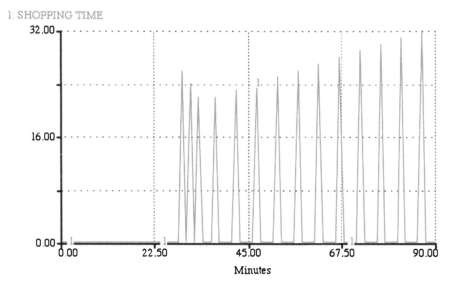

Fig. 4.6 Shopping times

Assume you are the manager of a store of this type, but you are lucky to have two
service centers and two cash registers. There are three employees in your store, each of
whom is qualified to work at the SERVICE CENTER and at the CASH REGISTER.
An employee can be at only one place at one point in time. It is your task to decide

when to have both service centers open and when to have both cash registers open. Find a work schedule that minimizes the mean shopping time over the course of an 8 h workday. Then, assume that you need to give your employees a set of breaks from their work; for example, two 15 min breaks each in the morning and afternoon, and a 1 h lunch break. How does this influence the "optimum" work schedule?

4.3 Store Model Equations

```
CASH_REGISTER(t) = CASH_REGISTER(t - dt) + (PAY - LEAVE) * dt
INIT CASH_REGISTER = 1
      COOK TIME = 2
      CAPACITY = 1
      FILL TIME = ∞
INFLOWS:
PAY = QUEUE TO OVEN FLOW
OUTFLOWS:
LEAVE = CONTENTS OF OVEN AFTER COOK TIME, ZERO OTHERWISE

CHECK_OUT_LINE(t) = CHECK_OUT_LINE(t - dt) + (MOVE_TO_CHECK_OUT
- PAY) * dt
INIT CHECK_OUT_LINE = 8
INFLOWS:

MOVE_TO_CHECK_OUT = CONVEYOR OUTFLOW
OUTFLOWS:
PAY = QUEUE TO OVEN FLOW

SERVICE_CENTER(t) = SERVICE_CENTER(t - dt) + (GET_SERVICE -
LEAVE_SERVICE_CENTER) * dt
INIT SERVICE_CENTER = 1
      COOK TIME = 5
      CAPACITY = 1
      FILL TIME = ∞
INFLOWS:
GET_SERVICE = QUEUE TO OVEN FLOW
OUTFLOWS:
LEAVE_SERVICE_CENTER = CONTENTS OF OVEN AFTER COOK TIME,
ZERO OTHERWISE

SERVICE_QUEUE(t) = SERVICE_QUEUE(t - dt) + (ENTER - GET_SERVICE) *
dt
INIT SERVICE_QUEUE = 2
INFLOWS:
ENTER = PULSE(1,3,4)  {People per Minute}
OUTFLOWS:
GET_SERVICE = QUEUE TO OVEN FLOW
```

```
TRANSFER(t) = TRANSFER(t - dt) + (LEAVE_SERVICE_CENTER -
MOVE_TO_CHECK_OUT) * dt
INIT TRANSFER = 0
        TRANSIT TIME = 1
        INFLOW LIMIT = ∞
        CAPACITY = ∞
INFLOWS:
LEAVE_SERVICE_CENTER = CONTENTS OF OVEN AFTER COOK TIME,
ZERO OTHERWISE
OUTFLOWS:
MOVE_TO_CHECK_OUT = CONVEYOR OUTFLOW
MEAN_SHOPPING_TIME = CTMEAN(LEAVE,1)
SHOPPING_TIME = CYCLETIME(LEAVE,1)
```

4.4 Optimizing Traffic Flow

The model of the previous section can be used to optimize the schedules of workers in a company that provides services, such as a grocery store or chemical lab, or to schedule workers or the use of materials in discrete production processes, such as assembly lines. The optimal schedule for the use of workers is often one that minimizes the number of people that are at the production facility at one point in time—full-time workers are used to meet the base demand and part-time workers augment them at peak times. We will encounter an example of the need for part-time workers in Chap. 10.

In some industries, it is customary to have a pool of employees on stand-by to meet unanticipated bottlenecks in the ability to produce a good or supply a service. A case in point is the use of on-call medical doctors and airline pilots. Similarly, optimization of material use in discrete production processes frequently results in only small inventories of input materials at the production facility. Therefore, to ensure that the system reliably meets demand even though it has only few workers or little excess materials on site requires efficient transportation of workers or materials to its site. Let us therefore investigate some features of transportation systems in more detail and model traffic flow for a simple commuter system to find the best schedule. Again, the ideas presented here can easily be used to model the transport of materials, for example, with freight trains or trucks.

Assume a simple commuter rail system connecting four stations. Passengers arrive at STATION 2 and STATION 3, and all leave at STATION 4. No passengers board at STATION 1 and STATION 4. STATION 1 is the shunting station from which trains successively leave to pick up passengers at STATION 2 and STATION 3. There are a constant number of passengers arriving at stations 2 and 3. The rate of passenger arrival at STATION 2 is 4 passengers per minute and that for STATION 3 is 5 passengers per minute. When a train arrives in the station and no train is in front of it, it stops for an average of 6 s per passenger before it can move on to the next station.

Trains pull out from STATION 1 into STATION 2 at fixed intervals. The rate at which trains pull out from STATION 1, PULLOUT, is specified by the built-in PULSE function:

$$PULLOUT = PULSE(1, DT, FREQ). \tag{4.4}$$

Each pulse releases 1 train (the first number in parentheses) into STATION 1. The first occurrence of that pulse is in time period DT. Thus, if DT = 1, then the first train leaves STATION 1 in period 1. The frequency at which trains depart STATION 1, FREQ, is a parameter and is specified exogenously to the model.

STATION 1 contains initially ten trains. There is no limit on the maximum number of trains that can be in STATION 1. However, no train can pull out from STATION 1 to STATION 2 if there is no train in STATION 1.

No more than 1 train can reach a station at a time. Trains leaving a station move through queues to reach subsequent stations. The travel time of a train, called TRAVEL in our model, depends on its speed from station to station (assumed to be a uniform 30 miles per hour) and the distance between stations (2.5 miles between stations 1 and 2; 2 miles between stations 2 and 3; 2 miles between stations 3 and 4; and 6.5 miles between stations 4 and 1). Additionally, the travel time depends on the time passengers at each station need to board the train. This DWELL time is the product of the BOARD TIME (6 s per passenger) and the total number of passengers that accumulated at a station between train arrivals.

Each station is modeled as an oven. Its initial value is 0 and its capacity is 1; that is, no more than 1 train can reach that station at a time. Trains leaving the station move through a queue to reach the next station. Queues prevent later trains to pass those that left a station earlier. The conveyor TRAIN 1 moves trains from STATION 1 into a queue TO 2. The transfer time depends on the distance from STATION 1 to STATION 2 and train speed. Once a train arrives in STATION 4, all passengers leave (at an average of 6 s per passenger) and the empty train returns to STATION 1 via a conveyor, TRAIN 4, with a transfer time dependent on distance and train speed.

Be careful in specifying the units in your model. Convert, for example, seconds and hours into minutes and run your model for the time scale of minutes. The part of our model that captures the movement of trains is shown in Fig. 4.7.

The number of passengers departing from a station depends on the accumulation of passengers at that station and whether a train has just pulled into the station. For example, to calculate the number of passengers departing STATION 2, we use the DELAY built-in function and a variable STATE OF S2 that indicates whether a train is in STATION 2. DELAY(variable,DT,0) specifies the value of a variable delayed by one DT and assigns a value of 0 to the value of the state variable at time $t = 0$. Thus, DELAY(STATE OF S2, 1, 0) is equal to 1 if a train has arrived in STATION 2 in the previous minute.

The STATE OF S2 variable is specified with the built-in function OSTATE as

$$STATE\ OF\ S2 = OSTATE(STATION\ 2), \tag{4.5}$$

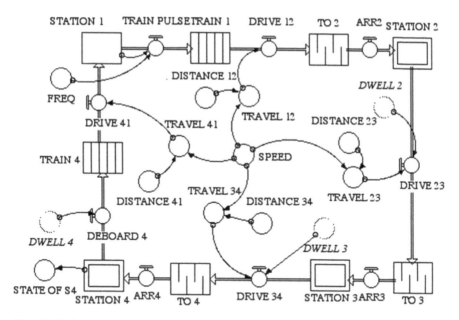

Fig. 4.7 Train movement

which gives the state of the oven STATION 2, and assumes 0 if the oven is empty, 1 if it is full, and 2 if it is emptying.

If a train just pulled into the station and no train was in the station during the previous minute, then people who arrived at the station since the departure of the last train will depart now; that is,

$$
\begin{aligned}
\text{PASS BOARD 2} = \text{IF(STATE OF S2} \\
= 1)\text{AND}((\text{DELAY(STATE OF S2}, 1, 0)) \\
\neq 1)\text{THEN PASSENGERS 2 ELSE 0} \qquad (4.6)
\end{aligned}
$$

Our model must keep track of the trains moving from station to station and the number of passengers on each train (Fig. 4.8). The latter information is required for the calculation of the time trains stay in STATION 4 before they move back to STATION 1 and has been used as a ghost in the module shown in Fig. 4.7 that captures train movement.

To calculate the time after which trains leave STATION 4, the number of passengers on each train must be known. When the passengers board each train (when PASS BOARD 2 or 3 >0), they are placed in a queue (PASS ON AT S2 or S3). The first number in each queue represents the number of passengers on the first train; the second number in each queue, the number of passengers on the second train; and so on. Thus when a train makes the nth complete trip to STATION 4 (STATE OF S4 becomes equal to 1), the nth element of the queues (PASS ON AT S2 or S3) represents the number of passengers on that train. This number is calculated in PASS ON TRAIN, using the built-in function QELEM. QELEM requires, as its first

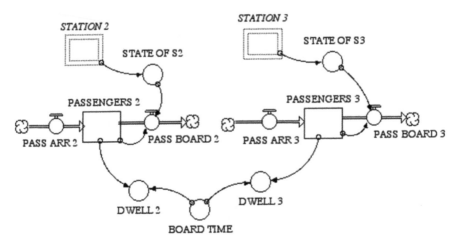

Fig. 4.8 Waiting passengers

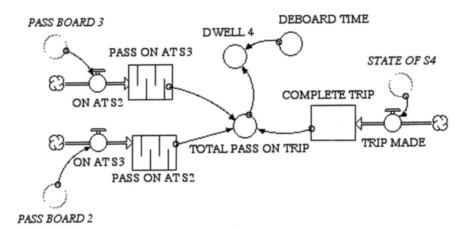

Fig. 4.9 Calculation of trips

input, the name of the queue and, as its second input, the element in that queue. In our model, we have

$$\text{TOTAL PASS ON TRIP} = \text{QELEM(PASS ON AT S2, COMPLETE TRIP} + 1) \\ + \text{QELEM(PASS ON AT S3, COMPLETE TRIP} + 1). \quad (4.7)$$

The 1 is added to the number of complete trips to offset the delay because it takes one time period to register trains arriving in STATION 4 as having made a complete trip (Fig. 4.9). From this number, DWELL 4 is calculated from the product of the number of passengers on the train and the DEBOARD TIME.

To optimize traffic flow, we may choose alternative frequencies, FREQ, at which trains pull into STATION 1. A number of different criteria may be chosen to assess optimality of traffic flow, such as the number of passengers waiting at each station or

Fig. 4.10 Average time
per trip

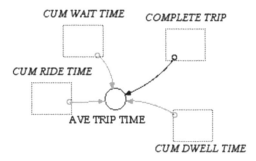

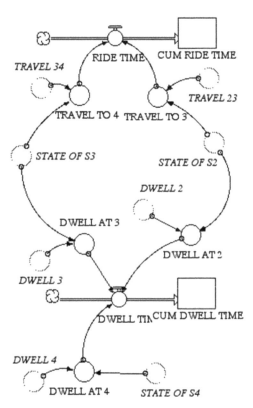

Fig. 4.11 Dwell and travel
times for trains

the mean travel time from STATION 2 to the destination STATION 4. A choice
must be made also with regard to the time frame over which we wish to optimize
traffic flow. In this model we choose to minimize the mean travel time from
STATION 1 to STATION 4 over the course of one half workday; that is, 240 min.

   To calculate the ride time (Fig. 4.10) we need to sum the dwell time for each
train (CUM DWELL TIME) and travel time for each train (CUM RIDE TIME),
both shown in Fig. 4.11, to the time each train waits in a queue (CUM WAIT
TIME), shown in Fig. 4.12. These cumulative times are then divided by the number
of complete trips that have been made to get the average ride time per train.

Fig. 4.12 Cumulative
waiting times

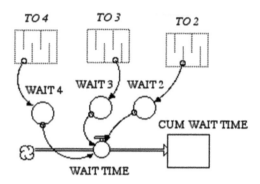

These cumulative times are calculated from the travel, dwell, and wait times for each train. Inflows to cumulative time reservoirs occur exactly when STATE OF S2, STATE OF S3, or STATE OF S4 is equal to 1, but not in the preceding time period.

Recognize how, in the long run, the average travel times per train arriving in STATION 3 become increasingly uniform. Given a time horizon of 4 h and one train pulling out to STATION 1 every 4 min, the average travel time per ride is approximately 1 h. Increasing the frequency at which trains arrive at STATION 2 may reduce the boarding time. However, for frequencies low enough, trains may get backed up in queues between stations, thereby increasing the travel time. Thus, it is not necessarily beneficial for traffic flow to have shorter spacing between the trains arriving at a station.

Decreasing the PULLOUT rates from 1 train every 4 min to 1 train every 7 min actually leads to an improvement in terms of the mean travel time. The five curves in Fig. 4.13 correspond to FREQ = 4, 7, 10, 13, 16 min. Increasing the spacing between trains to 10 min results in lower mean travel times, but increasing it further to 13 or 16 min leads to higher mean travel times.

Figure 4.14 shows the number of trains in STATION 1 for each of the five frequencies.

Try to find the schedule for the trains that minimizes the average trip time. We already know that it must be between FREQ = 10 and FREQ = 13.

You may want to add the time-stamp features discussed in the previous section and calculate the cycle time and standard deviation of the cycle time by using the built-in functions CYCLETIME and CTSTDDEV, respectively.

Extend the model to include changes in the rate of arrival of passengers, including features such as a well-pronounced rush hour. Find a new PULLOUT frequency that minimizes average travel time. In reality, we need to allow for the return of people to STATIONS 1 and 2 at the end of their workday (8 h). Make sure that the same number of people who left for work through each of the stations will arrive at that station at the end of the day.

In Chap. 2, we learned about ways of disaggregating stocks to answer questions that pertain to subsets of the overall system, such as the hiring and promotion of

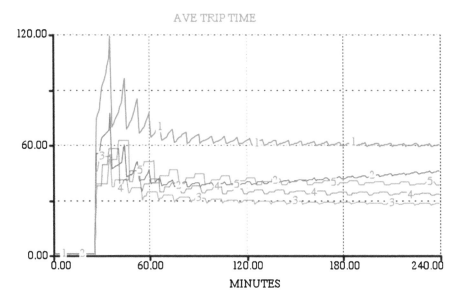

Fig. 4.13 Mean travel times for alternative departure frequencies

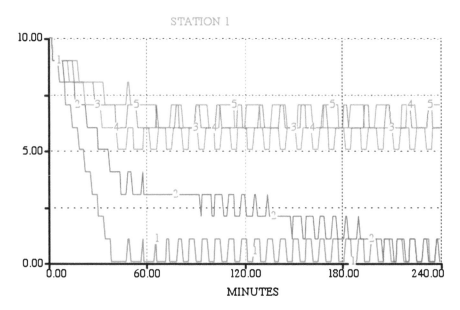

Fig. 4.14 Number of trains in station 1 for alternative departure frequencies

workers with different qualifications and genders in an individual company. In Chap. 3, we investigated the role of system boundaries in space and time for an assessment of changes in production technologies and the quality of resource endowments. In this chapter, we developed methods to optimize the performance of systems over space and time. All of the systems modeled so far had one important feature in common. There were no unforeseen changes in their environment: The company hiring new employees or promoting from within the firm had access to an infinite pool of qualified workers, and once hired no one left before retirement. Product life cycles, technologies, and changes in ore grades were all given and fixed for the firms of Chap. 3. Similarly, there was no random fluctuation in passenger arrivals at the train stations of Chap. 4, and all trains operated without breakdown and under constant weather conditions that enabled them to maintain a uniform mean speed between train stations. The real world is not that predictable. Unforeseen events may and will happen in a firm's environment. The effects of such events on a company's performance in the marketplace are modeled in the following chapter.

4.5   Traffic Flow Model Equations

COMPLETE_TRIP(t) = COMPLETE_TRIP(t - dt) + (TRIP_MADE) * dt
INIT COMPLETE_TRIP = 0 {Number of complete trips made}
INFLOWS:
TRIP_MADE = if (STATE_OF_S4=1) and ((delay(STATE_OF_S4,1,0))≠1) then 1 else 0 {Number of complete trips made per minute}

CUM_DWELL_TIME(t) = CUM_DWELL_TIME(t - dt) + (DWELL_TIME) * dt
INIT CUM_DWELL_TIME = 0 {Minutes; cumulative time spent boarding and de-boarding}
INFLOWS:
DWELL_TIME = DWELL_AT_2+DWELL_AT_3+DWELL_AT_4 {Minutes per minute; time spent boarding and deboarding passengers}

CUM_RIDE_TIME(t) = CUM_RIDE_TIME(t - dt) + (RIDE_TIME) * dt
INIT CUM_RIDE_TIME = 0 {Minutes; cumulative time of driving between stations 2 and 3 and between 3 and 4}
INFLOWS:
RIDE_TIME = TRAVEL_TO_3+TRAVEL_TO_4 {Minutes per minute; ride time to station 3 and to station 4}

CUM_WAIT_TIME(t) = CUM_WAIT_TIME(t - dt) + (WAIT_TIME) * dt
INIT CUM_WAIT_TIME = 0 {Minutes; cumulative time trains spend in queues waiting to enter the various stations}
INFLOWS:
WAIT_TIME = WAIT_2+WAIT_3+WAIT_4 {Minutes per minute; number of trains waiting one minute to enter the stations, implicit conversion of 1train in queue = 1 min of wait time}

PASSENGERS_2(t) = PASSENGERS_2(t - dt) + (PASS_ARR_2 - PASS_BOARD_2) * dt
INIT PASSENGERS_2 = 0 {Passengers waiting at STATION 2}
INFLOWS:
PASS_ARR_2 = 4 {Passengers arriving at STATION 2 per minute}
OUTFLOWS:

PASS_BOARD_2 = IF (STATE_OF_S2=1)
AND((DELAY(STATE_OF_S2,1,0))≠1) THEN PASSENGERS_2 ELSE 0 {Pas-
sengers per minute; number of passengers boarding the train at STATION 2}

PASSENGERS_3(t) = PASSENGERS_3(t - dt) + (PASS_ARR_3 -
PASS_BOARD_3) * dt
INIT PASSENGERS_3 = 0 {Passengers waiting at STATION 3}
INFLOWS:
PASS_ARR_3 = 5 {Passengers arriving at STATION 3 per minute}
OUTFLOWS:
PASS_BOARD_3 = IF (STATE_OF_S3=1) AND (DELAY(STATE_OF_S3,1,0)≠1)
THEN PASSENGERS_3 ELSE 0 {Passengers per minute; passengers boarding
the train at STATION 3}

PASS_ON_AT_S2(t) = PASS_ON_AT_S2(t - dt) + (ON_AT_S3) * dt
INIT PASS_ON_AT_S2 = 0 {Passengers; queue of passengers on each succes-
sive train}
INFLOWS:
ON_AT_S3 = if PASS_BOARD_2>0 then PASS_BOARD_2 else 0 {Passengers
per minute; number of passengers boarding train at station 2}

PASS_ON_AT_S3(t) = PASS_ON_AT_S3(t - dt) + (ON_AT_S2) * dt
INIT PASS_ON_AT_S3 = 0 {Passengers; queue of passengers on each succes-
sive train}
INFLOWS:
ON_AT_S2 = if PASS_BOARD_3>0 then PASS_BOARD_3 else 0 {Passengers
per minute; number of passengers boarding the train}

STATION_1(t) = STATION_1(t - dt) + (DRIVE_41 - TRAIN_PULSE) * dt
INIT STATION_1 = 10 {Trains; number of trains in STATION 1}
INFLOWS:
DRIVE_41 = CONVEYOR OUTFLOW
        TRANSIT TIME = TRAVEL_41 {Minutes per minute; travel time between
stations 4 and 1}
OUTFLOWS:
TRAIN_PULSE = IF STATION_1>0 THEN PULSE(1,DT,FREQ) ELSE 0 {Trains
per minute; pulse of train leaving STATION 1}

STATION_2(t) = STATION_2(t - dt) + (ARR2 - DRIVE_23) * dt
INIT STATION_2 = 0 {Trains; trains in STATION 2}
        COOK TIME = varies
        CAPACITY = 1
        FILL TIME = ∞
INFLOWS:
ARR2 = QUEUE TO OVEN FLOW
OUTFLOWS:

DRIVE_23 = CONTENTS OF OVEN AFTER COOK TIME, ZERO OTHERWISE
       COOK TIME = TRAVEL_23+DWELL_2 {Minutes per minute; total ride
time between stations 2 and 3}

STATION_3(t) = STATION_3(t - dt) + (ARR3 - DRIVE_34) * dt
INIT STATION_3 = 0 {Trains; trains in STATION 3}
       COOK TIME = varies
       CAPACITY = 1
       FILL TIME = ∞
INFLOWS:
ARR3 = QUEUE TO OVEN FLOW
OUTFLOWS:
DRIVE_34 = CONTENTS OF OVEN AFTER COOK TIME, ZERO OTHERWISE
       COOK TIME = DWELL_3+TRAVEL_34 {Minutes per minute; ride time be-
tween stations 3 and 4}

STATION_4(t) = STATION_4(t - dt) + (ARR4 - DEBOARD_4) * dt
INIT STATION_4 = 0 {Trains; trains in STATION 4}
       COOK TIME = varies
       CAPACITY = 1
       FILL TIME = ∞
INFLOWS:
ARR4 = QUEUE TO OVEN FLOW
OUTFLOWS:
DEBOARD_4 = CONTENTS OF OVEN AFTER COOK TIME, ZERO
OTHERWISE
       COOK TIME = DWELL_4 {Minutes per minute; time it takes for passen-
gers to deboard at STATION 4}

TO_2(t) = TO_2(t - dt) + (DRIVE_12 - ARR2) * dt
INIT TO_2 = 0 {Trains; queue of trains waiting to enter STATION 2}
INFLOWS:
DRIVE_12 = CONVEYOR OUTFLOW
       TRANSIT TIME = TRAVEL_12 {Trains per minute; total travel time be-
tween stations 1 and 2}
OUTFLOWS:
ARR2 = QUEUE TO OVEN FLOW

TO_3(t) = TO_3(t - dt) + (DRIVE_23 - ARR3) * dt
INIT TO_3 = 0 {Trains; queue of trains waiting to enter STATION 3}
INFLOWS:
DRIVE_23 = CONTENTS OF OVEN AFTER COOK TIME, ZERO OTHERWISE
       COOK TIME = TRAVEL_23+DWELL_2 {Minutes per minute; total ride
time between stations 2 and 3}
OUTFLOWS:
ARR3 = QUEUE TO OVEN FLOW

TO_4(t) = TO_4(t - dt) + (DRIVE_34 - ARR4) * dt
INIT TO_4 = 0 {Trains; queue of trains waiting to enter STATION 4}
INFLOWS:
DRIVE_34 = CONTENTS OF OVEN AFTER COOK TIME, ZERO OTHERWISE
        COOK TIME = DWELL_3+TRAVEL_34 {Minutes per minute; ride time be-
tween stations 3 and 4}
OUTFLOWS:
ARR4 = QUEUE TO OVEN FLOW
TRAIN_1(t) = TRAIN_1(t - dt) + (TRAIN_PULSE - DRIVE_12) * dt

INIT TRAIN_1 = 0 {Trains; conveyor of trains between stations 1 and 2}
        TRANSIT TIME = varies
        INFLOW LIMIT = ∞
        CAPACITY = ∞
INFLOWS:
TRAIN_PULSE = IF STATION_1>0 THEN PULSE(1,DT,FREQ) ELSE 0 {Trains
per minute; pulse of train leaving STATION 1}
OUTFLOWS:
DRIVE_12 = CONVEYOR OUTFLOW
        TRANSIT TIME = TRAVEL_12 {Trains per minute; total travel time be-
tween stations 1 and 2}

TRAIN_4(t) = TRAIN_4(t - dt) + (DEBOARD_4 - DRIVE_41) * dt
INIT TRAIN_4 = 0 {Trains, conveyor of train between stations 4 and 1}
        TRANSIT TIME = varies
        INFLOW LIMIT = ∞
        CAPACITY = ∞
INFLOWS:
DEBOARD_4 = CONTENTS OF OVEN AFTER COOK TIME, ZERO
OTHERWISE
        COOK TIME = DWELL_4 {Minutes per minute; time it takes for passen-
gers to deboard at STATION 4}
OUTFLOWS:
DRIVE_41 = CONVEYOR OUTFLOW
        TRANSIT TIME = TRAVEL_41 {Minutes per minute; travel time between
stations 4 and 1}
AVE_TRIP_TIME = IF COMPLETE_TRIP≠0 THEN
((CUM_WAIT_TIME+CUM_RIDE_TIME+CUM_DWELL_TIME)/COMPLETE_TRI
P) ELSE 0 {Minutes per trip; average time of a trip between waiting to enter
STATION 2 and leaving STATION 4}

BOARD_TIME = .1 {Minutes per passenger boarding}
DEBOARD_TIME = .1 {Minutes per passenger; deboarding time for each pas-
senger}
DISTANCE_12 = 2.5 {Miles; distance between stations 1 and 2}

DISTANCE_23 = 2 {Miles; distance between stations 2 and 3}
DISTANCE_34 = 2 {Miles; distance between stations 3 and 4}
DISTANCE_41 = 6.5 {Miles; distance between stations 4 and 1}
DWELL_2 = BOARD_TIME*PASSENGERS_2 {Minutes; boarding time at STATION 2}
DWELL_3 = BOARD_TIME*PASSENGERS_3 {Minutes; boarding time at STATION 3}
DWELL_4 = DEBOARD_TIME*TOTAL_PASS_ON_TRIP {Minutes; deboarding time at STATION 4}
DWELL_AT_2 = IF STATE_OF_S2=1 AND ((DELAY(STATE_OF_S2,1,0))≠1) THEN DWELL_2 ELSE 0 {Minutes; time each train spends boarding passengers at STATION 2}
DWELL_AT_3 = IF (STATE_OF_S3=1) AND ((DELAY(STATE_OF_S3,1,0))≠1) THEN DWELL_3 ELSE 0 {Minutes; time each train spends boarding passengers at STATION 3}
DWELL_AT_4 = IF STATE_OF_S4=1 AND ((DELAY(STATE_OF_S4,1,0))≠1) THEN DWELL_4 ELSE 0 {Minutes; time spent deboarding passengers at STATION 4}
FREQ = 7 {Minutes; frequency of trains leaving STATION 1}
SPEED = 0.5 {Miles per minute; speed of the train}
STATE_OF_S2 = OSTATE(STATION_2) {unitless; gives the state of oven STATION 2, 0=empty, 1=full, 2=emptying}
STATE_OF_S3 = OSTATE(STATION_3) {unitless; gives the state of station 3, 0=empty, 1=full, 2=emptying}
STATE_OF_S4 = OSTATE(STATION_4) {status of station 4, 0=empty, 1=full,2=emptying}
TOTAL_PASS_ON_TRIP = QELEM(PASS_ON_AT_S2,COMPLETE_TRIP+1)+QELEM(PASS_ON_AT_S3, COMPLETE_TRIP+1) {Passengers; number of passengers on each train}
TRAVEL_12 = DISTANCE_12/SPEED {Minutes; travel time between stations 1 and 2}
TRAVEL_23 = DISTANCE_23/SPEED {Minutes; travel time between stations 2 and 3}
TRAVEL_34 = DISTANCE_34/SPEED {Minutes; travel time between stations 3 and 4}
TRAVEL_41 = DISTANCE_41/SPEED {Minutes; travel time between stations 4 and 1}
TRAVEL_TO_3 = IF (STATE_OF_S2=1) AND ((DELAY(STATE_OF_S2,1,0))≠1) THEN TRAVEL_23 ELSE 0 {Minutes; the amount of time each train spends driving to STATION 3 from STATION 2}
TRAVEL_TO_4 = IF (STATE_OF_S3=1) AND ((DELAY(STATE_OF_S3,1,0))≠1) THEN TRAVEL_34 ELSE 0 {Minutes; the time each train spend driving to STATION 4}
WAIT_2 = QLEN(TO_2) {Trains; number of trains waiting to enter STATION 2 each minute}
WAIT_3 = QLEN(TO_3) {Trains, number of trains waiting to enter STATION 3 each minute}
WAIT_4 = QLEN(TO_4) {Trains; the number of trains waiting to enter STATION 4}

Chapter 5
Positive Feedback in the Economy

There are in fact four very significant stumbling-blocks in the way of grasping the truth, which hinder every man however learned, and scarcely allow anyone to win a clear title to wisdom, namely, the example of weak and unworthy authority, longstanding custom, the feeling of the ignorant crowd, and the hiding of our own ignorance while making a display of our apparent knowledge. Every man is involved in these things, every rank is affected. For each person, in whatever walk of life, both in application to study and in all forms of occupation, arrives at the same conclusion by the three worst arguments, namely this is a pattern set by our elder, this the custom, this is the popular belief: therefore it should be upheld.

Roger Bacon, *Opus Majus*, 13th century, I.1.

5.1 Feedback in the Economy

The models we built in the preceding chapters sketched the behavior of dynamic systems in a simplified, mechanical way. First, the relationships among components of the system were defined and the initial conditions specified within particular bounds, then the dynamics of the entire system were hypothesized and the model was run. Running the same model over and over again resulted in the same dynamics. Real systems, however, do not behave that well, that regularly. Some random element can change the direction of a system's behavior and its ultimate state. Things can become particularly "messy," if even very small, random influences on the system behavior change the strength of positive and negative feedback processes, possibly leading the system to entirely different states.

A save-disabled version of STELLA® and the computer models of this book are available at www.iseesystems.com/modelingeconomicsystems.

M. Ruth and B. Hannon, *Modeling Dynamic Economic Systems*,
Modeling Dynamic Systems, DOI 10.1007/978-1-4614-2209-9_5,
© Springer Science+Business Media, LLC 2012

The effect of positive and negative feedback processes on dynamic systems was discussed in Chap. 1. You may recall that negative feedback processes lead a system into a state of equilibrium, and positive feedback leads away from equilibrium by reinforcing a given tendency of a system. Acting alone, positive feedback can lead a system into a new, unpredictable equilibrium state.

5.2   Positive Feedback

Because of the tendency to maintain equilibrium states, researchers have focused on negative feedback processes. Breaking with this tradition, economist Brian Arthur (1990) wrote of the vital influence of positive feedback on economic systems. Arthur saw positive feedback processes as driving forces in determining which of competing new technologies would dominate a market. Arthur used the development of the VCR industry as an example to explain the effect of positive feedback. The industry began with two different formats: VHS and Beta. Both formats entered the market at approximately the same time, with similar products, and at first seemed to share the market almost equally. Which technology would ultimately take over the market? No one knew for sure. And many people wanted to know. Manufacturers wanted to know whether to produce films in VHS or Beta. Retailers and video rental businesses wanted to know which format to stock. Consumers wanted to know which technology to buy. The market was unstable, as manufacturers, retailers, and consumers waited to see which format would win. A combination of corporate strategy and "luck" gave a slight lead to VHS. Building on this lead, VHS became the sole tape format on the market.

Because at the start markets are unstable, small, seemingly random increases to a new technology's market share can expand its growth exponentially. Indeed, in video technology, the growth of VHS drove Beta off the market. This is not a unique example. Other products and technologies competing against close, but mutually exclusive, rivals simultaneously entering a new market have had similar experiences.

To capture positive feedback in a market, assume two producers are in the race for market dominance. At the beginning of this "race," each produces the same output quantity. Furthermore, each of the producers is free to adjust output at each period of time. We denote output by the two producers as $Q1$ and $Q2$, respectively. From these we can calculate the producer's market share $F1$ and $F2$. We can also keep track of cumulative production, $Z1$ and $Z2$.

Let us assume that the more that has been produced by a particular producer, the more experience that producer has gained, and as a consequence, the lower its production cost per unit of the output. Lower cost translates into lower prices $P1$ and $P2$, which, in turn, increase the attractiveness of the product to the consumer. The relationship between cumulative production ($Z1$) and price ($P1$), and that between price ($P1$) and "attractiveness" ($A1$) of the output to consumers are modeled with the use of Figs. 5.1 and 5.2.

Fig. 5.1 Price as a graphical
function of cumulative
production

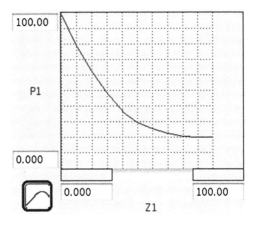

Fig. 5.2 Attractiveness
as a function of price

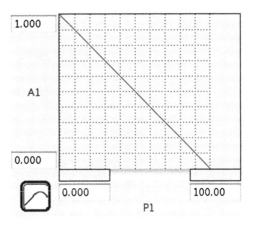

Let us assume that both producers have the same two curves. Also assume there are always unknown, random influences on the marketplace. These are captured by the variable RAND, defined as a random number varying between 0 and 1:

$$\text{RAND} = \text{RANDOM}(0, 1). \tag{5.1}$$

The choice on how much to expand production from one period to the next in the light of these random influences is assumed to be done in the following way. If the current market share of a producer exceeds the random number, then this is taken by the producer as a sign that things are going well for its product. Subsequently, production is increased above the previous level. If the market share does not

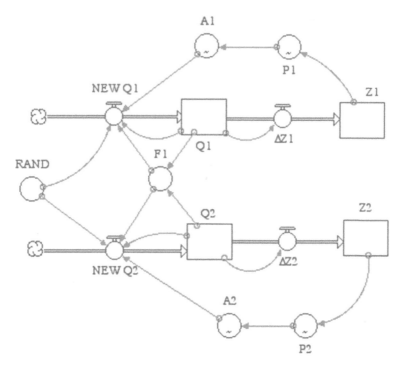

Fig. 5.3 Positive feedback model

exceed the random number, then production is held steady; that is, the new output level is equal to the previous output level. For producer 1 we have

$$\text{NEW } Q1 = \text{IF } F1 > \text{RAND THEN } Q1^*(1+A1) \text{ ELSE } Q1. \qquad (5.2)$$

Analogously,

$$\text{NEW } Q2 = \text{IF } F1 > \text{RAND THEN } Q2 \text{ ELSE } Q2^*(1+A2). \qquad (5.3)$$

These decision rules capture positive feedback. As the market share increases, it becomes more likely that the market share will exceed the random number, and consequently production is increased, speeding up the increase in cumulative production. As a result, price declines and attractiveness increases (Fig. 5.3).

The results of the model are shown for ten model runs in Fig. 5.4. Even though producer 1 always starts with a 50% share of the market and may be able to increase its market share early on, ultimate market dominance may not necessarily follow. However, once a significant share has been reached, it is almost impossible to lose the race for market dominance. Can you identify what approximately the critical share is? Run the model for many times and check the reliability of your answer.

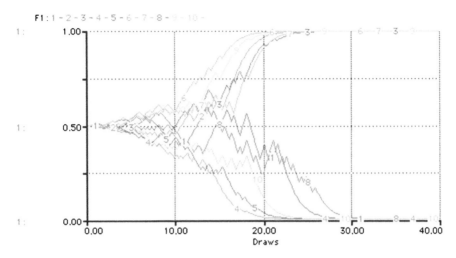

Fig. 5.4 Positive feedback model results

Modify the model to capture negative feedback. You can do that in many ways. For example, let the producers increase their production when they recognize that they fall back in the race or let the price increase with increased rates of production. Run the model many times. What do you find? Introduce a third producer into this model, and explore again the roles of positive and negative feedback for the system's dynamics.

We have now seen the effects of randomness on the market performance of a firm. Before we move on to investigate in more detail the behavior of firms in the marketplace, such as the optimum choice of input and output quantities, we need to learn how we can use STELLA to make a system move toward an optimum point as its dynamics unfold. This is the topic of the following chapter.

5.3 Positive Feedback Model Equations

```
Q1(t) = Q1(t - dt) + (NEW_Q1 - ΔZ1) * dt
INIT Q1 = 1
INFLOWS:
NEW_Q1 = IF F1 > RAND THEN Q1*(1+A1) ELSE Q1
OUTFLOWS:
ΔZ1 = Q1

Q2(t) = Q2(t - dt) + (NEW_Q2 - ΔZ2) * dt
INIT Q2 = 1
INFLOWS:
NEW_Q2 = IF F1 > RAND THEN Q2 ELSE Q2*(1+A2)
OUTFLOWS:
ΔZ2 = Q2
```

Z1(t) = Z1(t - dt) + (ΔZ1) * dt
INIT Z1 = 1
INFLOWS:
ΔZ1 = Q1

Z2(t) = Z2(t - dt) + (ΔZ2) * dt
INIT Z2 = 1
INFLOWS:
ΔZ2 = Q2

F1 = Q1/(Q1+Q2)
RAND = RANDOM(0,1)
A1 = GRAPH(P1)
(0.00, 1.00), (10.0, 0.9), (20.0, 0.8), (30.0, 0.7), (40.0, 0.6), (50.0, 0.5), (60.0, 0.4), (70.0, 0.3), (80.0, 0.2), (90.0, 0.1), (100, 0.00)
A2 = GRAPH(P2)
(0.00, 1.00), (10.0, 0.9), (20.0, 0.8), (30.0, 0.7), (40.0, 0.6), (50.0, 0.5), (60.0, 0.4), (70.0, 0.3), (80.0, 0.2), (90.0, 0.1), (100, 0.00)
P1 = GRAPH(Z1)
(0.00, 99.0), (10.0, 78.5), (20.0, 62.0), (30.0, 48.0), (40.0, 36.5), (50.0, 29.5), (60.0, 25.5), (70.0, 22.5), (80.0, 20.5), (90.0, 20.0), (100, 20.0)
P2 = GRAPH(Z2)
(0.00, 99.0), (10.0, 78.5), (20.0, 62.0), (30.0, 48.0), (40.0, 36.5), (50.0, 29.5), (60.0, 25.5), (70.0, 22.5), (80.0, 20.5), (90.0, 20.0), (100, 20.0)

Reference

Arthur B (1990) Feedbacks in the economy. Sci Am 262:92–99

Chapter 6
Derivatives and Lags

> *The order of historical events clearly shows the true position
> of the variational principle: It stands at the end of a long
> chain of reasoning as a satisfactory and beautiful condensa-
> tion of the results.*
>
> Max Born

6.1 Introduction

Integration is the process whereby a rate variable is added, in increments, to a stock,
an accumulated amount. Obviously, integration is necessary in solving rate or
differential equations, and STELLA is designed to perform this process. STELLA
can also do the opposite, differentiate, and find the rate of change of a variable or
slope of a curve. We would use this process to identify a variable's maximum or
minimum value.

Suppose we specified that a variable Y depends on another variable X, such as the
population depending on the number of births. Y is a function of X, and in general
this functional relationship may take on quite complicated forms. How can we use
STELLA to find the value of X that yields a maximum Y?

Take the example of an aquaculture firm that has special holding cages for adult
fish. The size of the cage is fixed and the firm manager wants to use the cage as
effectively as possible. As the manager of the firm, you may want to add only a
small number of fish to the empty cage to give each fish enough room. With this
strategy, you may need a lot of cages to keep all the fish. Could you have kept more
fish in the cage? Perhaps, but after a certain number of fish are added to the cage, the
fish population may suffer from crowding. Some of the smaller fish may die

A save-disabled version of STELLA® and the computer models of this book are available at
www.iseesystems.com/modelingeconomicsystems.

M. Ruth and B. Hannon, *Modeling Dynamic Economic Systems*,
Modeling Dynamic Systems, DOI 10.1007/978-1-4614-2209-9_6,
© Springer Science+Business Media, LLC 2012

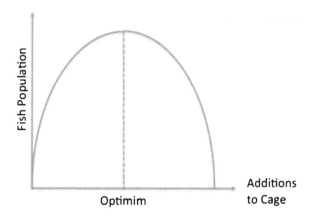

Fig. 6.1 Schematic fish
population dynamic

because they are not successful in competing for the food added to the cage, or they
may be prone to diseases that can spread quickly if the fish density in the cage is too
high. So your goal is to maximize the number of fish in a cage (variable Y) by
adjusting the number of fish that you add to the cage (variable X).

To find the optimum number of fish for each cage, we may set up the model to
plot the number of surviving fish as more and more fish are added to the cage
(Fig. 6.1). At first the fish population may increase, then reach a maximum, and as
more fish are added, they decrease in response to starvation or disease. As long as
additions of fish to the cage increase the size of the population, the slope of the
population curve is positive. At the point at which the maximum is reached, the
slope of the population curve is 0; to the right of this point, it is negative.

We can use STELLA to calculate the slope of the curve as fish are incrementally
added to the cage. STELLA provides a built-in list of derivative functions, which
we can use to find derivatives. Alternatively, we can take the value of a function,
subtract the value it held in the preceding time step (DT), and divide this difference
by DT. This gives us the rate of change, corrected for the size of DT, within a single
time period.

In Chap. 1, we briefly discussed how STELLA numerically determines the time
path of state variables. We used Euler's method, set in STELLA as the default
integration method, to numerically approximate that path. This method consists of
two steps. In the first step, the change in the state variable over the interval DT is
estimated from the flows F in time period t. The latter, in turn, are a function of the
stocks and possibly the values of other flows and converters at the beginning of the
period:

$$\Delta \text{state variable} = \text{DT}^*F(t, \text{ state variable, flows, converters}). \quad (6.1)$$

Stocks are then updated by adding to them the change in the state variable,
Δstate variable.

In the second step, new values for flows and converters are calculated as a function of the state variables, converters, and flows:

$$\text{flow} = F(\text{state variable, flows, converters}). \tag{6.2}$$

This is the simplest and fastest numerical solution method in STELLA.

Two other integration techniques are available and can be chosen in the TIME SPECS menu: Runge–Kutta-2 and Runge–Kutta-4. The second-order Runge–Kutta method consists of three steps to calculate the change in state variables before the new values for flows and converters are calculated. First, an estimate of the change in a state variable over the time period t is made from the flows F in that period:

$$F1 = \text{DT}^*F(t, \text{state variable, flows, converters}). \tag{6.3}$$

This is the same as with Euler's method. Then, a second estimate is calculated by projecting a DT ahead in the following way:

$$F2 = \text{DT}^*F(t + DT, \text{state variable} + F1, \text{flows, converters}). \tag{6.4}$$

These two estimates are combined in a third step to estimate the change in the state variable:

$$\Delta\text{state variable} = \text{DT}^*\frac{1}{2}(F1 + F2), \tag{6.5}$$

which is used to estimate the new state variable by adding Δstate variable to the value of the state variable that was present at the beginning of the time period.

The fourth-order Runge–Kutta method estimates changes in stocks by making even more forecasts of flows:

$$F1 = \text{DT}^*F(t, \text{state variable, flows, converters}), \tag{6.6}$$

$$F2 = \text{DT}^*F\left(t + \frac{DT}{2}, \text{state variable} + \frac{F1}{2}, \text{flows, converters}\right), \tag{6.7}$$

$$F3 = \text{DT}^*F\left(t + \frac{DT}{2}, \text{state variable} + \frac{F1}{2}, \text{flows, converters}\right), \tag{6.8}$$

$$F4 = \text{DT}^*F(t + DT, \text{state variable} + F3, \text{flows, converters}), \tag{6.9}$$

$$\text{state variables} = \frac{1}{6}(F1 + 2^*F2 + 2^*F3 + F4). \tag{6.10}$$

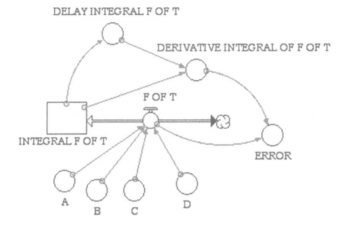

Fig. 6.2 Calculation of derivatives and integrals in STELLA

Runge–Kutta-4 is the slowest of the three numerical integration methods provided by STELLA and, as you will find, oftentimes the most accurate. But experiment with the three methods to see how your results are affected by the choice of method.

Now let us develop a model of integration over time, the calculation of a stock that increases stepwise over a specific period. This may also be called the integral of a function of time. The flow (F) steadily increases the stock as time goes on, so it is the function of time. For example,

$$F(T) = A + B^*\text{TIME}^{\wedge}C - \text{TIME}^{\wedge}D \tag{6.11}$$

with $A = 1, B = 10, C = 2$, and $D = 3$. Set up the model using a biflow for $F(T)$, because it will at times add to the integral and at times subtract from it (Fig. 6.2). We want the derivative of the integral. In STELLA, the derivative is the delayed value of the stock minus the value of the stock in time t, divided by DT. Because we use the numerical solution for the integral rather than the analytical one, we have a percentage error. This error can be calculated by comparing the value of the derivative we found using STELLA with the actual value of the flow. Run the model with Euler's method. See how smaller increments of DT can reduce the error. Rerun the model with Runge–Kutta-2 and Runge–Kutta-4, starting with a DT of 1 and reducing the DT for consecutive runs. Compare your results of the different model runs. Which is the "appropriate" method to use? Which is a "small enough" DT? Next we will examine some examples and uses of derivatives and lags. Investigate the sensitivity of your results with regard to the numerical solution technique.

In our model, the error is rather small until INTEGRAL $F(T)$—the state variable—changes its sign (Fig. 6.3). After the ERROR peaks, it soon settles back to a value close to zero.

Return to the example of the aquaculture firm trying to find the optimum stock of fish in a cage. Set up the model to yield an increase in population size at low density

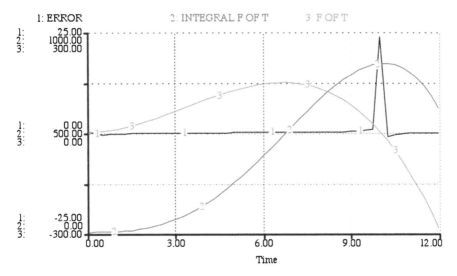

Fig. 6.3 Numerical results for the calculation of derivatives and integrals

of fish in the cage, and a decline in population at high densities. Find the maximum population size for a cage, running the model at various DT and with alternative integration methods.

6.2   Integration Model Equations

```
INTEGRAL_F_OF_T(t) = INTEGRAL_F_OF_T(t - dt) + (F_OF_T) * dt
INIT INTEGRAL_F_OF_T = 0
INFLOWS:
F_OF_T = TIME^2

DELAY_INTEGRAL_F_OF_T = DELAY(INTEGRAL_F_OF_T,DT)
DERIVATIVE_INTEGRAL_OF_F_OF_T = (INTEGRAL_F_OF_T -
DELAY_INTEGRAL_F_OF_T)/DT
ERROR = IF TIME > 0 THEN (DERIVATIVE_INTEGRAL_OF_F_OF_T-
F_OF_T)/F_OF_T
ELSE 0
```

6.3   Derivatives and Lags: Some Applications

6.3.1   Single-Output Firm

The derivative can prove a handy tool for finding the extremes of a variable's range. As we have seen in the previous section, STELLA has a built-in function, DELAY, which can compute the delayed (or lagged) value. This, in turn, can be used to determine the derivative as the difference between the actual and delayed values. We begin with an example that identifies the extreme ranges of a variable by using DELAY. A firm adjusts its output levels, in consecutive periods, to achieve a profit maximum. The SIGNAL assumes a value of 0 as long as the profit maximum has not been reached and a value of 1 to indicate the profit-maximizing output level.

For the model, assume that each period of time a new output level is chosen as

$$Q = 0.4^*\text{TIME}, \tag{6.12}$$

such that the cost of producing Q units of output is given by the cost function C as

$$C = 20^*Q^{\wedge}2 \tag{6.13}$$

and that prices P decrease with increasing output according to the demand function

$$P = 100 - 12^*Q. \tag{6.14}$$

The SIGNAL is defined to yield a value of 1, once the profit-maximizing output level is reached; that is,

$$\text{SIGNAL} = \text{IF } ((\text{CVP} - \text{DELAY}(\text{CVP}, \text{DT}) \leq 0))\text{AND}(\text{TIME}{>}1) \\ \text{THEN } 1 \text{ ELSE } 0, \tag{6.15}$$

with CVP as current value of profits:

$$\text{CVP} = P^*Q - C \tag{6.16}$$

and CCVP representing the cumulative current value of profits.

Nothing in the model (Fig. 6.4), however, prevents the firm from exceeding the profit-maximizing output level. As a result, with increased output the firm realizes losses (Fig. 6.5). Extend the preceding model to prevent the firm from surpassing its profit-maximizing output level. Be careful to avoid circularity in your model. One possible solution to this problem is shown in the following section.

6.3.2   Single-Output Firm

```
CCVP(t) = CCVP(t - dt) + (CVP) * dt
INIT CCVP = 0  {Dollars}
INFLOWS:
CVP = P*Q - 20*Q^2  {Dollars per Time Period}

P = 100-12*Q  {Dollars per Unit of Output}
Q = .4*TIME  {Units of Output per Time Period}
SIGNAL = IF ((CVP-DELAY(CVP,DT)) ≤ 0)) AND (TIME > 1) THEN 1 ELSE 0
{Dimensionless}
```

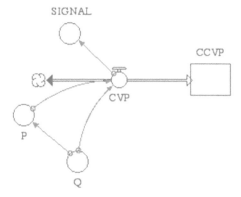

Fig. 6.4 Current value and cumulative value calculations

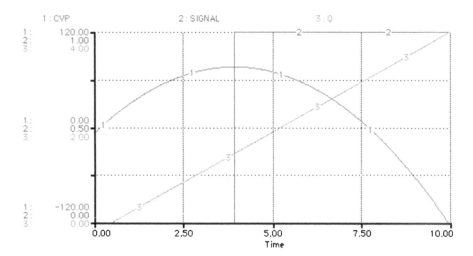

Fig. 6.5 Changes in current value profits (CVP) with changing output (Q)

6.3.3 Two-Output Firm

Let us extend the previous model to account for two different goods produced by the firm and sold on the market. This problem reveals a more complicated signal than the one just discussed. Here, we have products characterized by different cost functions and demand curves (Figs. 6.6 and 6.7). Cost functions for output $Q1$ and $Q2$ are, respectively,

$$C1 = 20^*Q1^\wedge2, \tag{6.17}$$

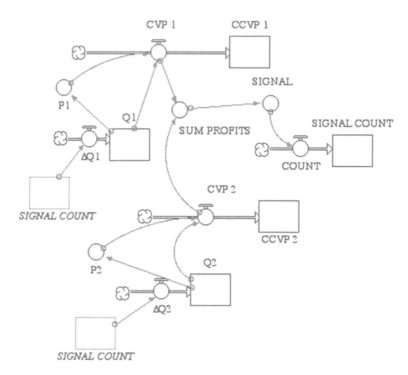

Fig. 6.6 Optimal production for the two-output firm

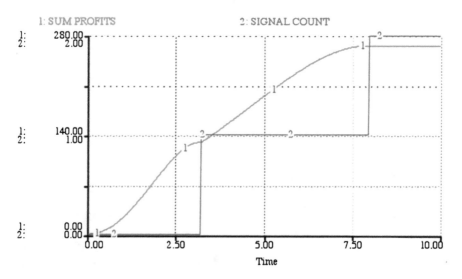

Fig. 6.7 Results for the two-output firm model

$$C2 = 60^*Q2^\wedge 2 \tag{6.18}$$

and the demand curves are given by

$$P1 = 100 - 12^*Q1, \tag{6.19}$$

$$P2 = 250 - 20^*Q2. \tag{6.20}$$

Assume additionally that adjustments in output levels are easier for output $Q1$ than output $Q2$:

$$\Delta Q1 = \text{IF SIGNAL\_COUNT} < 1 \text{ THEN } .4^*\text{TIME ELSE } 0 \tag{6.21}$$

$$\Delta Q2 = \text{IF SIGNAL\_COUNT} < 2 \text{ THEN } .05^*\text{TIME ELSE } 0. \tag{6.22}$$

The model enables us to find numerically when the peaks in the profits from each of the products occur. Additionally, the model maintains the output of each product at its profit-maximizing level. This is done by setting the change in output levels $\Delta Q1$ and $\Delta Q2$ equal to 0 once a peak in CVP 1 or CVP 2 has occurred. In this model, there are clearly two peaks, provided we started with small initial output levels.

The SUM PROFITS variable calculates the total profits in each period. These are used to generate a signal in the same way as we did before. Because there are multiple peaks in the SUM PROFITS function, we need to count peaks. This is done with the variable COUNT and the SIGNAL COUNT stock. COUNT generates a value based on the SIGNAL that is corrected for the length of DT. SIGNAL COUNT keeps track of the number of peaks that have occurred. Thus, this model can easily be expanded to accommodate multiple peaks and simulation runs with different time step lengths.

6.3.4   Two-Output Firm

```
CCVP_1(t) = CCVP_1(t - dt) + (CVP_1) * dt
INIT CCVP_1 = 0
INFLOWS:
CVP_1 = P1*Q1 - 20*Q1^2

CCVP_2 = CCVP_2(t - dt) + (CVP_2) * dt
INIT CCVP_2 = 0
INFLOWS:
CVP_2 = P2*Q2 - 60*Q2^2
```

Q1(t) = Q1(t - dt) + (ΔQ1) * dt
INIT Q1 = 0
INFLOWS:
ΔQ1 = IF SIGNAL_COUNT < 1 THEN .4*TIME ELSE 0 {Units of X per time
period}

Q2(t) = Q2(t - dt) + (ΔQ2) * dt
INIT Q2 = 0
INFLOWS:
ΔQ2 = IF SIGNAL_COUNT < 2 THEN .05*TIME ELSE 0

SIGNAL_COUNT(t) = SIGNAL_COUNT(t - dt) + (COUNT) * dt
INIT SIGNAL_COUNT = 0
INFLOWS:
COUNT = IF SIGNAL - DELAY(SIGNAL,DT) > 0 THEN 1/DT ELSE 0

P1 = 100-12*Q1
P2 = 250-20*Q2
SIGNAL = IF ((SUM_PROFITS-DELAY(SUM_PROFITS,DT) _ 0)) AND (TIME >=
1) THEN 1 ELSE 0
SUM_PROFITS = CVP_1+CVP_2

Part III
Microeconomic Models of Firms

Chapter 7
Introduction to Modeling Economic Processes

Perhaps it is the very simplicity of the thing which puts you at fault.

Edgar Allen Poe, "The Purloined Letter."

7.1 Core Principles of Economics

In the preceding part of this book we learned about the gradual development of a dynamic model, the role of system boundaries, models of discrete processes, multiple independent variables, randomness, feedback processes, and lags. These are the basic concepts used in all our models. For the development of models in the previous chapters, we combined our understanding of the system's driving forces with experiments on the computer. We will now begin to see the importance of blending analytical and numerical solutions of our models.

Analytical methods can make our models more efficient by speeding up the numerical techniques provided by STELLA and by increasing their accuracy. Analytics can give us equations that will describe not just optimal operating points but whole optimal trajectories through time—a very difficult problem to formulate numerically. However, analytical methods will soon carry us to a point at which their further use would be either inefficient or impossible—the optimal operating points and trajectories are too difficult for analytics to completely solve. For example, many of the nonlinear dynamic relationships that underlie our models have no analytical solutions. This is one reason why economic models typically have been developed to investigate the properties of an equilibrium situation, rather than the potentially complex path toward that equilibrium. By combining analytical and

A save-disabled version of STELLA® and the computer models of this book are available at www.iseesystems.com/modelingeconomicsystems.

M. Ruth and B. Hannon, *Modeling Dynamic Economic Systems*,
Modeling Dynamic Systems, DOI 10.1007/978-1-4614-2209-9_7,
© Springer Science+Business Media, LLC 2012

numerical techniques in the following chapters, we build on the strengths of both methods and arrive at a solution superior to using just one of the methods. The economic models can then be solved for potentially complex paths toward equilibrium.

Each academic specialty uses principles unique to that field. Economics builds on three principles: substitution, time value, and opportunity cost (for a detailed discussion see Ruth 1993). An example of substitution is when machines replace workers on a production line. Here, inputs such as the machines, the energy they consume, and the labor of technical workers increase as the production time and work hours of less-skilled (or differently skilled) workers decrease.

If you borrow $1,000 from a bank for 1 year, you will have to repay not only $1,000 but also a premium (interest on the loan) to the bank for foregoing other uses of that money for a year. Foregoing consumption can be expensive and risky. During the year when you borrow the money, a more profitable investment might emerge, and the bank would have to wait to use the money you borrowed. Also, you might not repay the loan or repay only part of it. The interest (the premium you pay in addition to the $1,000) reflects the time value of the loan.

An accountant who earns $50/h wants her living room repainted. She could do the job herself or hire a painter for $25/h. If she hires a painter, the accountant could accept extra work during those evenings when she would have been painting. The cost of those evening hours, the opportunity cost, is $25/h.

These principles—substitution, time value, and opportunity cost—are central in economics. In the following three chapters, we provide illustrations for each of these concepts.

At this point, we would like to refer you to the list of symbols and their explanation at the beginning of the book. We will make extensive use of the notation outlined there.

Reference

Ruth M (1993) Integrating economics, ecology, and thermodynamics. Kluwer Academic, Dordrecht

Chapter 8
Substitution of Inputs in Production

*Division of labor may have gone so far that many people—
men especially—feel disconnected and alienated from the
ultimate goals of their work. And that may partly be what
leads individuals to seek satisfaction in exaggerated
expressions of virtuosity ... Much professional culture may
be a disguise for this sense of emptiness, devoted as so much
of it is to building up a notion of the professional importance
of the profession.*

Arnold Pacey, *Culture of Technology*, 1983.

8.1 Trade-Off Possibility Frontiers

Firms in the real world use a number of different inputs to generate a desired output.
Even a process as simple as making some hand-made wooden toy requires labor,
wood, some kind of knife, and knowledge as its fundamental inputs. Of course,
some of these inputs can be substituted for each other, at least within some range.
For example, if you have a lot of experience in producing wooden toys, you may be
able to cut down on the time you spend per toy. Similarly, if everything else remains
equal and you work more slowly, and thus more carefully, you will take longer to
produce a product but you may be able to cut down on waste of the wood. You trade
off, or substitute, labor for the input material.

In this chapter we model a "trade-off possibility frontier" for two inputs at a
given output level. A trade-off possibility frontier is also called an *isoquant* and

A save-disabled version of STELLA® and the computer models of this book are available at
www.iseesystems.com/modelingeconomicsystems.

M. Ruth and B. Hannon, *Modeling Dynamic Economic Systems*,
Modeling Dynamic Systems, DOI 10.1007/978-1-4614-2209-9_8,
© Springer Science+Business Media, LLC 2012

Fig. 8.1 Specification of an
isoquant and cost function

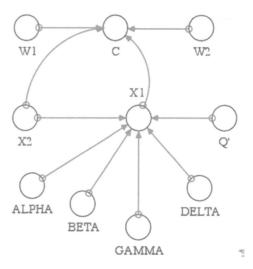

shows the minimum input combinations that just generate a desired output. Let us
assume that the production function for the firm is of the form

$$Q = \text{GAMMA} * (\text{ALPHA} * X1^{\text{BETA}} + (1 - \text{ALPHA}) * X2^{\text{BETA}})^{\text{DELTA/BETA}}. \quad (8.1)$$

where ALPHA, BETA, GAMMA, and DELTA are fixed parameters that establish
the "recipe" for producing output Q from inputs $X1$ and $X2$.

Since the isoquant provides the relationship between inputs $X1$ and $X2$ for a fixed
output level, we define such an output level as Q^o, insert Q^o into (8.1), and solve for
one of the inputs, such as $X1$. This gives us $X1$ as a function of $X2$, the parameters of
the production function, and the fixed output level:

$$X1 = \frac{1}{\text{ALPHA}} * \left[\left(\frac{Q^o}{\text{GAMMA}} \right)^{\text{BETA/DELTA}} - (1 - \text{ALPHA}) * X2^{\text{BETA}} \right]^{1/\text{BETA}}.$$

$$(8.2)$$

Suppose $W1$ and $W2$, the prices of $X1$ and $X2$, respectively, are given. The
following equation can be used to calculate the cost to a firm of producing output Q^o:

$$C = W1 * X1 + W2 * X2. \quad (8.3)$$

We can now build a model that determines an isoquant and cost function for an
output level of $Q^o = 4.0$ units (Fig. 8.1). The parameters are set as ALPHA $= 0.8$,
BETA $= 0.3$, GAMMA $= 1$, and DELTA $= 0.8$. Unit cost of inputs are $W1 = 2$
and $W2 = 1$.

The model is run for a fixed output level of $Q^o = 4.0$ units. Figures 8.2 and 8.3
illustrate the trade-off possibility frontier and cost functions generated by the model.

Fig. 8.2 Cost function

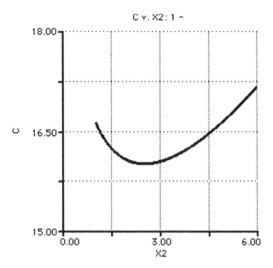

Fig. 8.3 Isoquant

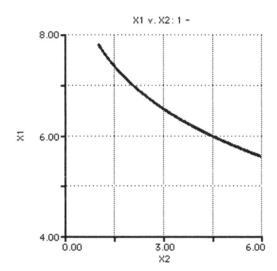

The minimum of the cost function is $16. In that minimum, the firm uses approximately 6.6 units of $X1$. Given the cost-minimizing $X1$, the optimal choice of $X2$ is approximately 2.5 units. The second graph shows the relationships between $X1$ and $X2$, the isoquant.

The production function of (8.1) is called a *constant elasticity of substitution production function*, or for short, CES production function. CES production functions are characterized by a constant percentage change in the ratio of inputs in response to a 1% change in the slope of the isoquant. Can you confirm this property for the function we used here? Model the isoquants for a production function of the type $Q = A*X1\^ALPHA*X2\^BETA$. What are the cost-minimizing input combinations for alternative, fixed output levels? How does the ratio of inputs $X1$ and $X2$ change for this production function as we move along the isoquant?

In the model of this section, we assumed that an output level was predetermined. Then the cost function's minimum was used to determine the firm's choice of a profit-maximizing input combination. We did not know, however, whether our choice of the output level Q^o did indeed maximize the profits of the firm. In practice, the profit-maximizing output level and cost-minimizing input combination are best determined simultaneously by a firm. The next section demonstrates how this can be done.

8.2 Trade-Off Possibilities Frontier Model Equations

```
ALPHA = .8
BETA = .3
C = W1*X1 + W2*X2  {Dollars}
DELTA = .8
GAMMA = 1
Q0 = 4 {Units of Output}
W1 = 2 {Dollars Per Units of X1}
W2 = 1  {Dollars Per Units of X2}
X1 = (((Q0/GAMMA)^(BETA/DELTA))/ALPHA - ((1-
ALPHA)/ALPHA)*X2^BETA)^(1/BETA)
X2 = .05*(TIME)+1  {Units of Input}
```

8.3 Profit Maximization with Several Inputs

The firm in our present model is searching for the profit-maximizing output level from its fixed input costs ($W1$, $W2$) and output price (P). We assume here that the production function bears the same form as in the last section:

$$Q = \text{GAMMA} * (\text{ALPHA} * X1^{\text{BETA}} + (1 - \text{ALPHA}) * X2^{\text{BETA}})^{\text{DELTA/BETA}}. \quad (8.4)$$

Once again, ALPHA, BETA, GAMMA, and DELTA are the parameters describing the relationship between inputs and output.

We want to find the mixture of $X1$ and $X2$ that minimizes the firm's costs while maximizing its profits. As presently formulated, this problem does not lend itself to a numerical solution using STELLA; first, we must perform some analysis. CVP is the current value of profit. It is defined as

$$\text{CVP} = P^*Q - W1^*X1 - W2^*X2. \quad (8.5)$$

We now draw briefly on some analytic reasoning to save ourselves an immense amount of extra programming. We will do this again, particularly in the chapters on optimal resource use through time. To find the profit-maximizing conditions, we take the partial derivatives of the CVP function relative to $X1$ and $X2$:

$$\frac{\partial \mathrm{CVP}}{\partial X_1} = P * \frac{\partial Q}{\partial X_1} - W_1 = 0, \tag{8.6}$$

$$\frac{\partial \mathrm{CVP}}{\partial X_2} = P * \frac{\partial Q}{\partial X_2} - W_2 = 0. \tag{8.7}$$

The partial derivatives are used to calculate the change in CVP in response to very small changes in inputs $X1$ and $X2$. Because we wish CVP not to change for marginal changes in inputs—that is, we want to hold CVP at its maximum—we set these derivatives to 0. If small changes in $X1$ or $X2$ would lead to increases in CVP—that is, the marginal derivatives are positive—we would not have chosen the profit-maximizing inputs. Similarly, for decreases in CVP in response to marginal changes in $X1$ or $X2$, we have surpassed the optimum input levels. Because both partial derivatives must be 0 at the same time, we combine these two equations to yield the optimum condition:

$$X2 = \left(\frac{W2}{W1} \frac{\mathrm{ALPHA}}{(1 - \mathrm{ALPHA})} \right)^{1/(\mathrm{BETA}-1)^*} X1. \tag{8.8}$$

Equation (8.8) establishes the relationship between the two inputs. After we have determined a profit-maximizing output level for one input, say, $X1$, it is not difficult to determine the optimum quantity for the other, $X2$. We then proceed as in the previous section, manipulating one input until we reach the profit maximum. Once again, $X1$ rises with a positive slope on the profit curve. The rise in $X1$ brings about a consequent rise in $X2$, which together cause Q to increase. The difference between revenue and ($P*Q$) and cost yields the profit based on a cost-minimizing mix of inputs. Although we used only two inputs here, any number of multiple inputs can be treated in the same way. We encourage you to set up the problem for the cases of more than two inputs.

The following STELLA diagram of Fig. 8.4 shows the two-input situation, assuming a small value for $X1$ at first. The profit-maximizing input level $X2$ that corresponds to our choice of $X1$ is determined by (8.8). From the levels of $X1$ and $X2$ we can calculate the firm's output (Q) and cost (C); hence, its profit. The profit just calculated may be less than maximum. Until the profit-maximizing output level is reached—that is, as long as inputs $X1$ and $X2$ can be increased to produce an increase in CVP, over the profits reached in the preceding small time step (DT)—we continue to increase $X1$, and consequently $X2$ (Fig. 8.5).

The profit-maximizing output level shown in this model, $Q = 4$, is the level chosen to model the isoquant earlier. We wanted to find the optimum (cost-minimizing) input mix that yields the profit-maximizing output level. The results of the new model and those of the model in the preceding section correspond well.

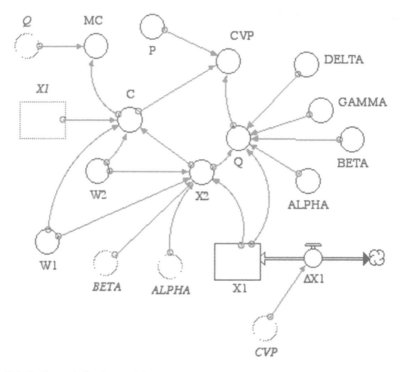

Fig. 8.4 Profit-maximization model

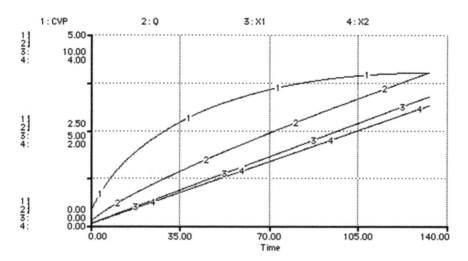

Fig. 8.5 Profit-maximization results

Figure 8.6 illustrates the relationship between marginal cost (MC), the change in cost that follows a change in Q, and price (P). We see that MC increases to the level of P when profit is maximum, just as predicted by microeconomic theory.

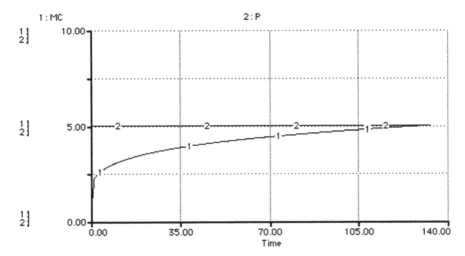

Fig. 8.6 Marginal cost and price

Of course, by combining this technique with methods used in the previous models, we can be more effective in finding a numerical solution using STELLA. Often, the most accurate and efficient solution to a problem requires the appropriate mix of analytic and numerical analysis.

Now that we have modeled substitution—a core concept of economics—we will turn our attention to the concept of time preference. This is the topic of the following chapter.

8.4 Competitive Firm with Substitution Model Equations

```
X1(t) = X1(t - dt) + (ΔX1) * dt
INIT X1 = .1  {Units of X1}
INFLOWS:
ΔX1 = IF (CVP-DELAY(CVP,DT,.1)) > 0 THEN .05 ELSE 0 {Units of X1 per Time
Period}

ALPHA = .8
BETA = .3
C = W1*X1+W2*X2  {Dollars}
CVP = P*Q-C  {Dollars}
DELTA = .8
GAMMA = 1
MC = (C-DELAY(C,dt,.24))/(Q-DELAY(Q,dt,.14))  {Dollars per Units of Q}
P = 5  {Dollars per Unit Q}
Q = GAMMA*(ALPHA*X1^BETA + (1-ALPHA)*X2^BETA)^(DELTA/BETA)  {Units
of Q}
W1 = 2  {Dollars per Unit X1}
W2 = 1  {Dollars per Unit X2}
X2 = ((W2/W1)*ALPHA/(1-ALPHA))^(1/(BETA-1)) * X1  {Units of X2}
```

Chapter 9
Time Value

We shall not cease from exploration
And the end of all exploring
Will be to arrive where we started
And know the place for the first time.

T. S. Eliot, 1943

9.1 Current and Present Value Calculation

In the previous chapter, we modeled a firm that applied a given technology to produce a desired output. We started the models under the assumption that the initial input quantities, and therefore the level of output, were too low. Then, we let input quantities and output adjust to yield the profit maximum. Such problems of profit-maximizing firms are typically solved in economics as static problems, but we pretended that adjustments took place over time. Let us now explicitly introduce time into decision-making by the firm.

With the following models, we try to answer the questions of how firms can compare profits that occur at different periods of time. Having a dollar today is surely different from having a dollar next year. In order to compare profits that accrue at different time periods, we may calculate their present value rather than their current value. In later chapters, we take up the issue of present vs. current value in more detail and provide the tools necessary to assess, for example, optimal resource use by firms over time.

Typically, firms value higher those profits that occur in the present or nearby future than profits in the distant future. There are many reasons for valuing profits differently based on their occurrence. For example, the timing and amount of profits may be

A save-disabled version of STELLA® and the computer models of this book are available at www.iseesystems.com/modelingeconomicsystems.

M. Ruth and B. Hannon, *Modeling Dynamic Economic Systems*,
Modeling Dynamic Systems, DOI 10.1007/978-1-4614-2209-9_9,
© Springer Science+Business Media, LLC 2012

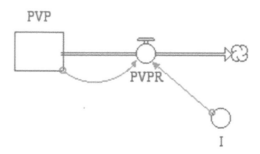

Fig. 9.1   Present value calculation

more uncertain for the more distant future. Furthermore, if profits accrue today, they may be reinvested to generate even higher profits in the future. Therefore, lower weights are assigned to future profits when firms evaluate a stream of profits over time.

The value of a fixed profit from today's perspective is the lower the further in the future that profit is earned. The percentage rate at which the present value of the profits, PVP, declines into the future is called the *discount rate*. It captures risk, uncertainty, and alternative investment possibilities. For practical purposes we may choose a fixed interest rate, I, as the discount rate.

In Fig. 9.1, the stock of PVP is drawn down by a given interest rate of 3%. The stock contains an initial value of $200. The outflow present value of profit rate, PVPR, deceases PVP over time. The outflow depends on the remaining stock and the interest rate. Because both PVP and the interest rate, I, influence the size of the outflow PVPR over time, both are connected to that outflow with information arrows.

The model shows the present value of $200 at various times, t, in the future. Running the model at a DT $= 0.25$, you can see from Fig. 9.2, for example, that the value of $200 for the decision-maker in the present is only approximately $110, 20 years into the future. Conversely, if the firm had invested $110 today at the given interest rate, it would own $200 in 20 years from now.

If we are interested in the amount received t periods from now for an investment of $200 at a given interest rate, we just need to reverse the direction of the flow in the STELLA model, thereby accumulating value. Such a calculation gives us the future value of our investment. In this case, the stock rises exponentially.

Often we do not invest a given amount of money in just one period but repeat the investment over a number of periods. Similarly, firms typically receive a stream of profits over time. In Fig. 9.3, the current value of an annual investment series of $200 is calculated and displayed in the graph. In our model, the steady investment stream combined with interest earned on the investment lead to an increase in cumulative current value of profits, CCVP, of close to $130,000 after 100 years (Fig. 9.4).

Investments may not be uniform over time. Similarly, streams of profits may follow cycles of high and low sales, and interest rates may fluctuate. For simplicity, we assume that there is a constant stream of annual payments, AP 2, of $100 and

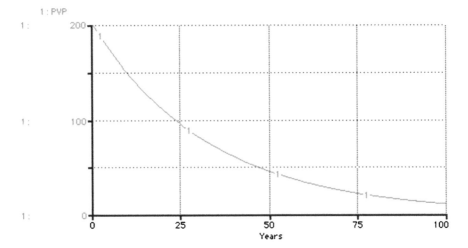

Fig. 9.2 Changes in present value through time

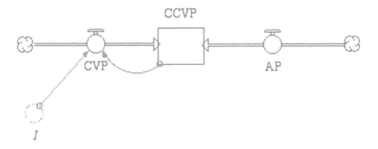

Fig. 9.3 Accumulation of profits

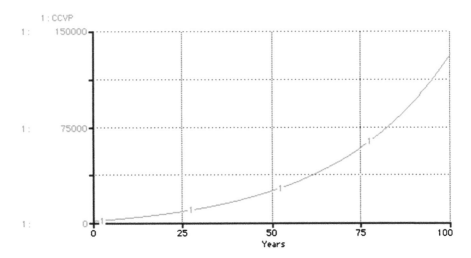

Fig. 9.4 Profit accumulation through time

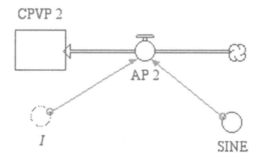

CPVP 2

AP 2

I   SINE

Fig. 9.5 Fluctuating income stream

that interest rates change in a cyclic pattern. Let us calculate the cumulative present value of our investments where the interest rate changes along a sine wave.

To specify the sine wave pattern for the interest rate, create a converter, call it SINE, and open it. Select SINWAVE from the list of built-in functions. SINWAVE requires the specification of an amplitude and period of the wave, each written within the parentheses and separated by a comma. If you choose an amplitude of 0.02 and a period of 50 years, then the sinewave has a maximum value of 0.02, a minimum of -0.02, and a mean of 0. The sinewave is repeated every 50 years. Use that specification of the sinewave. You should now have

$$SINE = SINWAVE(.02, 50) \tag{9.1}$$

as the function defining the variable SINE. Next, open the flow AP 2 and specify it as

$$AP\ 2 = 100^{*}EXP(-(I + SINE)^{*}TIME) \tag{9.2}$$

to generate a stream of annual payments of \$100 that is continuously discounted by a rate that fluctuates along a sine wave between 1 and 5%; that is, in the interval $I \pm 0.02$. The resulting cumulative present value is calculated and plotted in Fig. 9.5.

Introduce into this model variations in the investment stream. You can do this in different ways. Try specifying the investment stream to also follow a sine wave that shows declines in payments during the first 25 years and subsequent increases (Fig. 9.6). Alternatively, specify the investment streams with a graph to allow for rather erratic changes in annual payments. What are the implications of these erratic payments for the cumulative present value of profits in the long run? We will return to this issue of variable income streams combined with variable interest rates in Chap. 22.

In calculating the present value, one can use either $EXP(-I * TIME)$ or the factor $(1 + I * DT)^{\wedge}(-TIME/DT)$ as the analytical expression (with $^{\wedge}$ as the symbol for exponentiation). The first one is continuous (or infinitely often) negative

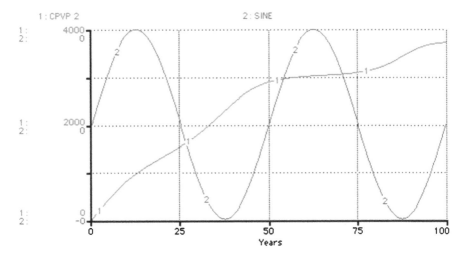

Fig. 9.6 Accumulation with fluctuating income stream

compounding, and the second is stepwise (DT) negative compounding. Run the
model subsequently with the two discounting methods for DT = 1 and note the
difference. The graph in the second case rises to the same present value of some
string of future values but it rises more quickly. Change DT to smaller values and run
the model repeatedly. You should find the curves get closer and closer. The exponential discounting is just getting more accurate but the stepwise discounting is
actually changing to more and more frequent negative compounding, a conceptually
different result. As DT goes to 0, the discount factor $(1 + I * DT)^{\wedge}(-TIME/DT)$ is
$EXP(-I*TIME)$ as you can easily show in STELLA.

9.1.1   Time Value Model Equations

```
CCVP(t) = CCVP(t - dt) + (CVP + AP) * dt
INIT CCVP = 200  {Dollars}
INFLOWS:
CVP = I*CCVP  {Dollars per Time Period}
AP = 200  {Dollars per Year}
CPVP_2(t) = CPVP_2(t - dt) + (AP_2) * dt
INIT CPVP_2 = 0  {Dollars}
INFLOWS:
AP_2 = 100*EXP(-(I+SINE)*TIME)  {Dollars per Time Period}
PVP(t) = PVP(t - dt) + (- PVPR) * dt
INIT PVP = 200  {Dollars}
OUTFLOWS:
PVPR = PVP*I  {Dollars per Year}
I = .03  {Dollars per Dollar per Year}
SINE = SINWAVE(.02,50)
```

9.2 Cost–Benefit Analysis

Now that we learned how to calculate the present value of future payments, let us employ the method of discounting revenues and cost to decide among alternative investment options. Such comparisons are regularly made in policy decision-making when society needs to decide, for example, whether to build a dam for flood control or a new highway to improve the accessibility of a region. Let us use this method of comparing discounted cost and benefits of a project for an individual company.

Your company is ready to commit $5 million in profits from last year's operations to a possible combination of proposed company projects or a bank account. The market interest rate for money in the bank account is 10%. Abstract away from future taxes.

There are three projects, A, B, and C, from which the firm can choose. Each project is characterized by a distinct construction, maintenance, and revenue stream. For project A, these streams are specified in Figs. 9.7–9.9.

Construction, maintenance, and revenue streams for projects B and C are in the model that accompanies this book. Which of the investment options should your firm choose?

To answer the question, form a discounted net present value, PVP, of each investment then accumulate it for a choice of the return rate (Fig. 9.10). Make sure that you allow your stock of the cumulative net present value, CPVP, from which you subtract the PVP, to be able to become negative. Run the model first with a low rate of return; note the result. Then, increase the rate of return for subsequent runs and compare the model results. Find that return rate which causes CPVP of the project to be exactly 0. This is defined as the internal rate of return for the project. Compare this rate to those of the other investment options. We have chosen a 50-year lifetime for each project, after which time we assume it is abandoned.

Figure 9.11 shows the cumulative net present value of project A over the 50-year time horizon for rates of returns set at 5, 6, 7, 8, and 9%, respectively. Plot the results

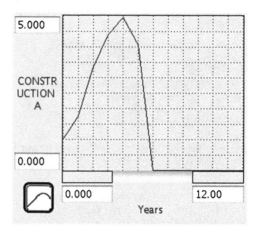

Fig. 9.7 Construction cost profile for project A

Fig. 9.8 Maintenance cost
profile for project A

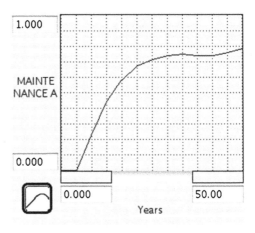

Fig. 9.9 Revenue cost profile
for project A

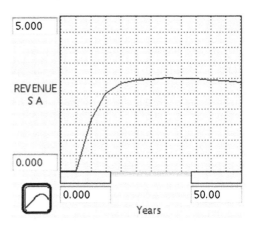

Fig. 9.10 Cost benefit model

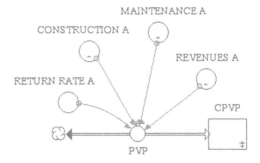

in a table to find the rate of return that comes closest to CPVP = 0 in year 50.
The analysis of project A should give a rate of return of approximately 7.89%.
Find the other rates of return and determine which will draw the profit funds. Then
invest the money in the market while drawing it down over the construction years to

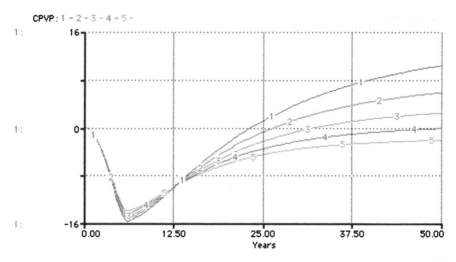

Fig. 9.11 Cumulative present values of profits for different rates of return

finance the winning project(s). What is the present value of the invested profit over the 50-year time span?

Earlier we noted that cost–benefit analysis is a method frequently employed both at the level of an individual firm and for policy decision -making on projects such as dams for flood control or construction of highways. Assume that a particular project, such as a dam for flood control, may be carried out by either an individual firm or a government agency. Under each scenario, construction, maintenance, and revenue streams are the same. What may compel an individual company to *not* carry out the project even though it may be beneficial from a societal perspective and perhaps be carried out by the government agency?

Of course, the discount rate is a crucial determinant for the choice of a project, and we may expect that individual firms have shorter planning horizons than society as a whole. The firm may go out of business if it does not generate sufficient profits. Its decisions should be made in the interest of the owners of the firm, all of whom have a finite life expectancy and wish to see their investments pay off soon. Government, in contrast, may be around for a long time to come and is supposed to represent the interests not just of the current generation but also of that generation's children and grandchildren. As a consequence, the firm's discount rate may be higher than the government's, and therefore, its choice of an interest rate to which to compare the rate of return of the project may be different.

Which discount or interest rate to apply in cost–benefit analysis is a hotly debated issue. For a detailed discussion, see, for example, Lind et al. (1982). The model presented here should help you investigate the implications of alternative assumptions about the discount rate. We will return to this issue later, when we determine the rates at which nonrenewable and renewable resources should be extracted from the environment to maximize cumulative present value of profits.

9.3 Cost–Benefit Model Equations

CPVP(t) = CPVP(t - dt) + (PVP) * dt
INIT CPVP = 0 {Dollars}
INFLOWS:
PVP = exp(-RETURN_RATE_A*time)*(REVENUES_A - MAINTENANCE_A -
CONSTRUCTION_A) {Dollars per Time Period}
CONSTRUCTION_A = GRAPH(time)
(0.00, 1.00), (1.00, 1.75), (2.00, 3.35), (3.00, 4.38), (4.00, 4.95), (5.00, 4.05),
(6.00, 0.00), (7.00, 0.00), (8.00, 0.00), (9.00, 0.00), (10.0, 0.00), (11.0, 0.00),
(12.0, 0.00)
CONSTRUCTION_B = GRAPH(time)
(0.00, 1.00), (1.00, 1.75), (2.00, 2.88), (3.00, 3.52), (4.00, 3.90), (5.00, 3.90),
(6.00, 3.08), (7.00, 2.08), (8.00, 0.775), (9.00, 0.00), (10.0, 0.00), (11.0, 0.00),
(12.0, 0.00)
CONSTRUCTION_C = GRAPH(time)
(0.00, 1.00), (1.00, 1.75), (2.00, 2.23), (3.00, 2.70), (4.00, 2.98), (5.00, 2.42),
(6.00, 0.00), (7.00, 0.00), (8.00, 0.00), (9.00, 0.00), (10.0, 0.00), (11.0, 0.00),
(12.0, 0.00)
MAINTENANCE_A = GRAPH(time)
(0.00, 0.00), (4.17, 0.00), (8.33, 0.235), (12.5, 0.44), (16.7, 0.585), (20.8, 0.675),
(25.0, 0.715), (29.2, 0.74), (33.3, 0.75), (37.5, 0.74), (41.7, 0.74), (45.8, 0.76),
(50.0, 0.785)
MAINTENANCE_B = GRAPH(time)
(0.00, 0.00), (4.17, 0.00), (8.33, 0.00), (12.5, 0.265), (16.7, 0.34), (20.8, 0.385),
(25.0, 0.39), (29.2, 0.39), (33.3, 0.39), (37.5, 0.395), (41.7, 0.405), (45.8, 0.43),
(50.0, 0.44)
MAINTENANCE_C = GRAPH(time)
(0.00, 0.00), (4.17, 0.00), (8.33, 0.41), (12.5, 0.685), (16.7, 0.74), (20.8, 0.76),
(25.0, 0.77), (29.2, 0.79), (33.3, 0.805), (37.5, 0.82), (41.7, 0.84), (45.8, 0.86),
(50.0, 0.89)
RETURN_RATE_A = .0789
REVENUES_A = GRAPH(time)
(0.00, 0.00), (4.17, 0.00), (8.33, 1.65), (12.5, 2.52), (16.7, 2.85), (20.8, 2.95),
(25.0, 2.98), (29.2, 3.02), (33.3, 3.00), (37.5, 3.00), (41.7, 2.95), (45.8, 2.92),
(50.0, 2.88)
REVENUES_B = GRAPH(time)
(0.00, 0.00), (4.17, 0.00), (8.33, 0.00), (12.5, 2.85), (16.7, 3.77), (20.8, 4.25),
(25.0, 4.28), (29.2, 4.28), (33.3, 4.28), (37.5, 4.30), (41.7, 4.30), (45.8, 4.08),
(50.0, 2.88)
REVENUES_C = GRAPH(time)
(0.00, 0.00), (4.17, 0.00), (8.33, 1.65), (12.5, 2.52), (16.7, 3.42), (20.8, 3.88),
(25.0, 4.00), (29.2, 4.08), (33.3, 4.08), (37.5, 4.10), (41.7, 4.12), (45.8, 4.12),
(50.0, 4.12)

Reference

Lind RC, Arrow KJ, Corey GR, Dasgupta P, Sen AK, Stauffer T, Stiglitz JE, Stockfisch JA,
     Wilson R (1982) Discounting for time and risk in energy policy. Resources for the Future,
     Washington

Chapter 10
Opportunity Cost

It seems to me simplicity is about the most difficult thing to achieve in scholarship and writing. How difficult is clarity of thought, and yet it is only as thought becomes clear that simplicity is possible. When we see a writer belaboring an idea we may be sure that the idea is belaboring him. This is proved by the general fact that the lectures of a young college assistant instructor, freshly graduated with high honors, are generally abstruse and involved, and true simplicity of thought and ease of expression are to be found only in the words of the older professor. When a young professor does not talk in pedantic language, he is positively brilliant, and much may be expected of him.

Lin Yutang, *The Importance of Living*, 1937

10.1 Managing an Inventory

In this chapter we turn to the third core concept of economics—the concept of opportunity cost. Opportunity cost is the utility or profits foregone by choosing one alternative over another. Whether consciously or not, every day we make decisions by comparing opportunity cost. If the pleasure of going to the movies exceeds that of staying at home and reading a book, you will likely find yourself going to the movies.

Similarly, companies must compare alternative actions with each other and decide which to take. A rational strategy is to choose the action with the lowest opportunity cost. For example, a firm may decide on keeping some of its products in stock to be able to meet demand when demand, and thus price, is high. However,

A save-disabled version of STELLA® and the computer models of this book are available at www.iseesystems.com/modelingeconomicsystems.

M. Ruth and B. Hannon, *Modeling Dynamic Economic Systems*,
Modeling Dynamic Systems, DOI 10.1007/978-1-4614-2209-9_10,
© Springer Science+Business Media, LLC 2012

Fig. 10.1 Inventory

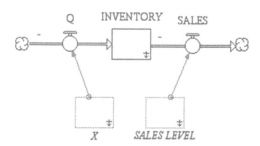

keeping an inventory is costly not just because goods have to be stored. Inventory cost also arises from the loss of income that could have been generated if the goods had been sold. In this chapter we calculate the opportunity cost associated with holding an inventory. For a simpler version of this model, see Hannon and Ruth (2001).

Let us assume that there is only one input, X, into the production process. That input generates Q units of output according to the following production function:

$$Q = 3^*X^{\wedge}0.3 \tag{10.1}$$

The output Q temporarily becomes part of the INVENTORY. The sales of goods from the inventory are equal to the sales level at the current period. The part of the STELLA model that captures the production–inventory–sales relationship is shown in Fig. 10.1.

Set DT $= 1/8$. Assume that the firm's decision on the level of production in a given period depends on the change in inventory. If the inventory declined since the last period, input X into the production process is reduced to reduce output. If the inventory increased since the last period, then production is decreased via a reduction in X, and the inventory is restocked. The adjustments in inputs, ΔX, in response to changes in inventory, ΔINVENTORY, are given in Fig. 10.2.

The choice of X by the firm leads to production cost, C, which is determined in the following way. Think of X as a labor input. Assume that the firm is committed, through contracts, to compensate workers for 40 h of work a week, even if the workers work less. The wage rate based on the 40-h workweek is fixed at $W = 1.0$. Overtime must be paid at higher rates. For the first 5 h of overtime the wage rate is $W = 1.5$, and thereafter $W = 2$. The conditional statement that captures this pay schedule is

$$\begin{aligned} W = &\text{IF } X > 40 \text{ AND } X \leq 45 \text{ THEN } 1.5 \text{ ELSE IF } X \\ &> 45 \text{ THEN } 2 \text{ ELSE } 1 \end{aligned} \tag{10.2}$$

and the corresponding production cost is

Fig. 10.2 Adjustment in inputs

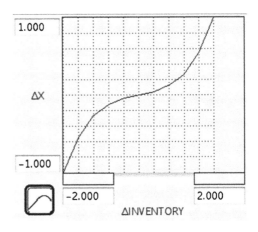

Fig. 10.3 Changes in input dependent on inventory

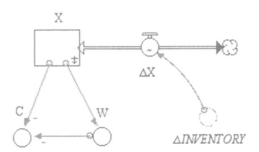

$$C = \text{IF } X \le 40 \text{ THEN } 40^*W \text{ ELSE } W^*X. \tag{10.3}$$

These relationships between inventory, inputs, wages, and production cost are given in Fig. 10.3.

Changes in inventory are based on an exogenously determined desired inventory, the actual size of the inventory, and a "damping factor." The DAMPING FACTOR, in turn, is calculated by comparing the current difference between the desired and actual inventory with that difference one time step earlier. To make this comparison meaningful, we need to correct for possible differences in the sign. Such a correction is made by taking the square of the difference between the desired and actual inventory.

$$\text{DAMPING FACTOR} = \text{IF}((\text{DESIRED INVENTORY}$$
$$- \text{INVENTORY})^{\wedge}2 > \text{DELAY}((\text{DESIRED INVENTORY}$$
$$- \text{INVENTORY})^{\wedge}2, \text{DT}, 0.1))\text{THEN } 8 \text{ ELSE } 1$$

$$\tag{10.4}$$

Fig. 10.4 Changes
in inventory

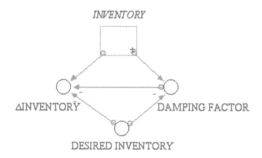

The resulting DAMPING FACTOR is used to calculate the change in inventory (Fig. 10.4):

$$\Delta INVENTORY = DAMPING\ FACTOR^*(DESIRED\ INVENTORY$$
$$-\ INVENTORY)/INVENTORY \qquad (10.5)$$

As we stated already, production costs are only part of the costs that need to be considered by the firm in its decision-making process. The opportunity cost of holding an inventory is a second determinant for the choice of appropriate levels of production and inventory. The opportunity cost of the inventory is the foregone sales. Therefore, to calculate these opportunity cost we must multiply the inventory by the market price and the rate of interest for the time step. Multiplying by the interest rate for the time step is necessary to calculate the amount of money that could have been earned had the inventory been sold and the profits from it invested. This procedure takes into account the concept of time preference discussed in the previous chapter.

For our model, the market price of the good produced by the firm is given by the following demand curve:

$$P = 15 - 0.01^*SALES \qquad (10.6)$$

Given the price, the opportunity cost of the inventory, and the cost of production, we can calculate the current value of profits, CVP, as shown in Fig. 10.5.

Changes in the level of sales, $\Delta SALES$, are assumed to be driven by changes in the current value of profits, ΔCVP. The relationship between $\Delta SALES$ and ΔCVP is defined by the graph in Fig. 10.6, and the corresponding STELLA module is shown in Fig. 10.7.

The firm adjusts its inventory so long until the change in profit becomes 0. The corresponding inventory, sales, output, and price are shown in Fig. 10.8. After significant initial adjustments long-run equilibria for each of the state variables are reached.

The CVP, cost of holding the inventory, and producing output Q are shown next, followed by a graph of input levels, wage rates, and production costs. In the

Fig. 10.5 Current value of profit

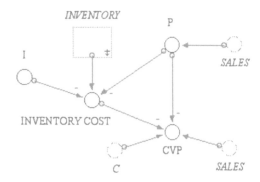

Fig. 10.6 Changes in sales

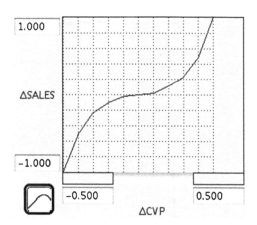

Fig. 10.7 Sales level

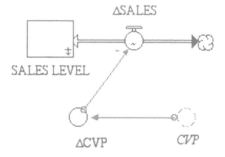

equilibrium, X exceeds 60 units and the firm must pay the wage rate $W = 2.0$. The cost of the inventory is approximately \$18, and CVP is approximately \$110 (Figs. 10.9 and 10.10).

The worker of this model will have to contribute more than 60 h per week to the production process, so that the profit-maximizing output level can be achieved.

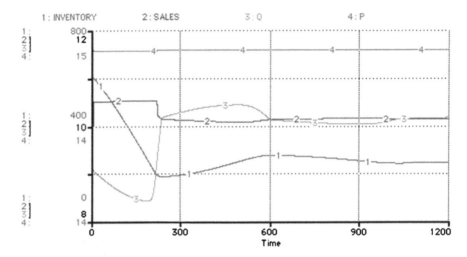

Fig. 10.8 Equilibrium outcomes

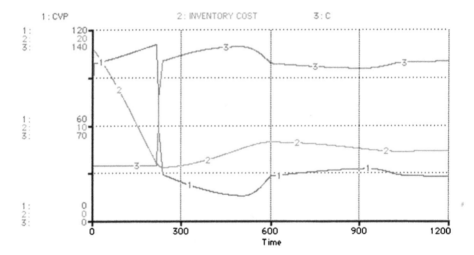

Fig. 10.9 Current value profits, cost of inventory, and production cost

This is obviously an undesirable situation—from the perspectives of both the worker and the firm that must pay the high wage rate for overtime. Recall our discussion in Chap. 4, where we indicated that under a variety of scenarios it may be beneficial for firms to hire part-time workers in addition to full-time workers. How would hiring a part-time worker affect the optimal inventory and the maximum CVP? Assume a different pay schedule for that worker and make appropriate assumptions on that worker's availability. Give the workers a significant pay raise over the wage rates that are assumed in our hypothetical firm. How high can W be so that the firm does not run out of business?

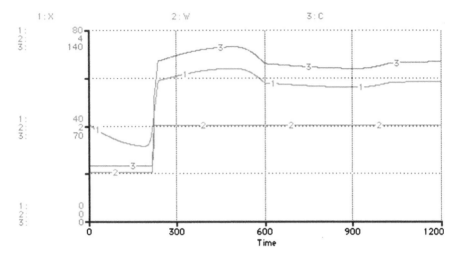

Fig. 10.10 Inputs, wage rate, and production cost

Change the specification of the damping factor for subsequent runs of the model. Recognize that, if you try to do without this factor in your model, you may run into stability problems. Can you find alternative specifications for ΔX that avoid making use of the DAMPING FACTOR, and in the calculation of ΔINVENTORY, yet achieve a profit maximum while reaching an INVENTORY at the desired level of 250 units?

The firm of this model maximizes its current value profit while it maintains inventory. Can you find a way to also optimize inventory at the same time?

10.2   Opportunity Cost Model Equations

```
INVENTORY(t) = INVENTORY(t - dt) + (Q - SALES) * dt
INIT INVENTORY =  600  {Units of the Product}
INFLOWS:
Q = 3*X^0.3  {Units of the Product per Time Period}
OUTFLOWS:
SALES =  SALES_LEVEL  {Units of the Product per Time Period}
SALES_LEVEL(t) = SALES_LEVEL(t - dt) + (ΔSALES) * dt
INIT SALES_LEVEL = 10  {Units of the Product}
INFLOWS:
ΔSALES = GRAPH(ΔCVP)
(-0.5, -1.00), (-0.4, -0.51), (-0.3, -0.23), (-0.2, -0.1), (-0.1, -0.02), (0.00, 0.00),
(0.1, 0.02), (0.2, 0.11), (0.3, 0.23), (0.4, 0.48), (0.5, 1.00)
X(t) = X(t - dt) + (ΔX) * dt
INIT X =  40  {Units of the Input per Time Period}
```

INFLOWS:
ΔX = GRAPH(ΔINVENTORY)
(-2.00, -1.00), (-1.60, -0.55), (-1.20, -0.26), (-0.8, -0.12), (-0.4, -0.04), (0.00, 0.00),
(0.4, 0.04), (0.8, 0.12), (1.20, 0.27), (1.60, 0.55), (2.00, 1.00)
C = IF X <= 40 THEN 40*W ELSE W * X {Dollars per Time Period}
CVP = P * SALES - C - INVENTORY_COST {Dollars per Time Period}
DAMPING_FACTOR = IF ((DESIRED_INVENTORY - INVENTORY)^2 >
    DELAY((DESIRED_INVENTORY - INVENTORY)^2,DT,0.1)) THEN 8
    ELSE 1 {The square removes the sign effect of the difference. 1/Time Period}
DESIRED_INVENTORY = 250 {Units of the Product}
I = 0.002
{$ Interest per $ Principal per Time Period}
INVENTORY_COST = I * P * INVENTORY {Dollars per Time Period}
P = 15 - 0.01 * SALES {Dollars per Unit of the Product}
W = IF X>40 AND X<=45 THEN 1.5 ELSE IF X > 45 THEN 2 ELSE 1 {Dollars
per Unit of the Input}
ΔCVP = (CVP - DELAY (CVP,DT,0.1)) / SQRT (CVP ^ 2) {Dollars per Time Pe-
riod}
ΔINVENTORY = DAMPING_FACTOR * (DESIRED_INVENTORY -
INVENTORY) / INVENTORY {Units of the Product per Time Period}

Reference

Hannon B, Ruth M (2001) Dynamic modeling, 2nd edn. Springer, New York

Chapter 11
The Profit-Maximizing Competitive Firm

In questions of Science the authority of a thousand is not worth the humble reasoning of a single individual.

Galileo Galilei

11.1 Optimizing Behavior of the Competitive Firm

In the previous chapter, we illustrated the use of core concepts of economics. Now we turn our attention to a series of models to illustrate frequently made assumptions on the behavior of firms. For example, a standard assumption is that firms expand the level of their output until they maximize the level of their profits. We made use of this assumption in Chaps. 6 and 8.

Another important assumption is that firms have perfect information about all current and future features of the economy and environment that are necessary for their decision-making. With the exception of Chap. 5, we have not yet explicitly modeled the market within which a firm makes its decisions. One crucial determinant for a firm's decision is the signals that it receives from the market. An assumption standard in microeconomic theory is that all relevant information for decision-making is subsumed in the price of a good or service. As prices increase, producers are compelled to enlarge their production, thus driving down the price. As prices decline, consumers are willing to buy more of a product, thus stimulating the price to increase. The decisions of producers and consumers are coordinated in the market, with the price as the signal that achieves that coordination. Of course, each consumer may have a different willingness to pay for the goods that are offered in the market; and if there is a shortage, potential buyers may try to out-bid their competitors.

A save-disabled version of STELLA® and the computer models of this book are available at www.iseesystems.com/modelingeconomicsystems.

M. Ruth and B. Hannon, *Modeling Dynamic Economic Systems*,
Modeling Dynamic Systems, DOI 10.1007/978-1-4614-2209-9_11,
© Springer Science+Business Media, LLC 2012

We will discuss such a model in more detail in Chap. 15. Similarly, adjustments of producers to price changes may not be instantaneous. We model that case in Chap. 32.

Before we return to these important issues, let us build on the assumptions of profit maximization and perfect information to determine the optimal size of a firm in a competitive market. This is the topic of this chapter.

Assume that you are the manager of a small firm that produces wooden toys. Your firm is one of many small firms that all try to sell identical output and compete with each other. Neither your firm nor any of your competitors individually has sufficient power to influence the price of hand-made wooden toys. Similarly, none of you can individually exert any influence on the suppliers of wood to lower the price of your essential input. As a result, both the price of wooden toys and the wood that you use to produce them are given to you and beyond your influence. How many toys should you produce in your firm? Answering this question requires that we identify the goal of your firm. One reasonable goal may be to maximize profits from producing and selling wooden toys. Another goal may be to rapidly expand your operation from an initial size to a desired one. Of course, both goals are related to each other. If your firm is small yet could increase its profits to a maximum by increasing its output, it should do so. The rate of expansion is likely closely related to the profits you make in a year. The higher are the profits, the more money you can set aside for future expansions and the larger loans you may be able to receive from a bank to support the expansion of your firm. Alternative assumptions about the optimal behavior of firms abound and may even be more appropriate: Satisficing behavior—meeting preset goals in a multiyear plan is sufficient indication of appropriate size and expansion rate—and expanding market share, to name but two.

Let us model your firm under the assumptions that it is small, does not influence the price of its inputs or output, and can expand faster the larger are its profits. For simplicity let us also assume that you have perfect information about all relevant market characteristics, such as the prices of inputs and outputs, and about the technology of producing wooden toys.

At a fixed price P, the revenues R from selling a quantity Q of wooden toys on the competitive market are

$$R = P * Q. \tag{11.1}$$

The technology used to produce Q units of toys from X units of wood is given by the following production function:

$$Q = A * X^\wedge \text{ALPHA}, \tag{11.2}$$

where $0 < \text{ALPHA} < 1$ ensures a declining rate of growth of Q as X increases. The wood, in turn, needs to be purchased at a unit cost W. Purchase of X units of wood results in total cost

$$C = W * X. \tag{11.3}$$

At total cost C and revenues R, the current value of profits is

$$\mathrm{CVP} = R - C. \tag{11.4}$$

This is the standard formulation for profit that we already introduced briefly in Chap. 6. In that chapter we showed how STELLA can be used to solve differential equations. The differential equation of interest to you as the manager of the firm is the equation that relates changes in profits to changes in output. This change in profit in response to a change in output, in turn, will determine the rate at which you can expand production. To calculate the change in profits as the level of output changes, we take the first partial derivative of the profit function (11.3) with respect to Q:

$$\frac{\partial CVP}{\partial Q} = \frac{\partial R}{\partial Q} - \frac{\partial C}{\partial Q} = \mathrm{MR} - \mathrm{MC} = P - \mathrm{MC}. \tag{11.5}$$

If you start out with a sufficiently small output, slight increases in output are likely to increase your revenues faster than you drive up the cost. As a result,

$$\frac{\partial CVP}{\partial Q} > 0. \tag{11.6}$$

Once your firm reached a *critical* size, even a very small increase in output may lead to an increase in revenues that is smaller than the increase in cost. Consequently,

$$\frac{\partial CVP}{\partial Q} < 0. \tag{11.7}$$

Thus, there is an output level at which profits cannot be increased further by marginal increases in output and

$$\frac{\partial CVP}{\partial Q} = 0. \tag{11.8}$$

This is the profit-maximizing amount of wooden toys that your firm should produce. Let us take the first partial derivative of the profit function (11.3) with respect to Q and set it equal to 0 to identify this profit-maximizing output level. Once we have identified this level, we will attend to the second issue of adjusting the expansion of your company in response to the change in its profits.

The first partial derivative of the profits with respect to Q is given in (11.5). Setting it equal to 0 yields the condition that in the optimum:

$$P = MC. \tag{11.9}$$

The second-order conditions for this and all the following problem sets are fulfilled.

Fig. 11.1 Profit-maximizing
output

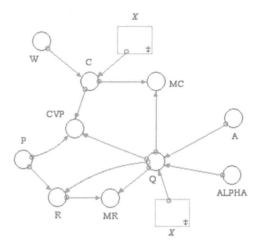

The following STELLA model (Fig. 11.1) is set up to calculate revenues R, cost C, profits CVP, and marginal cost MC. Each of those is determined by the choice of our output level Q. Outputs, in turn, are determined by our choice of input quantities X, given the production function (11.2). How should you choose the input quantities such that you achieve your goal of maximizing profits?

As the company is small, slight increases in inputs will result in disproportionate increases in outputs and profits. Large profits, CVP, should enable you to quickly expand your operation and to purchase a larger amount of wood in the next period. As long as there is an increase in profits, the firm should expand production—controlled through an appropriate choice of input quantities. In our model the change in profits is defined as

$$\Delta CVP = CVP - DELAY(CVP, DT, .1) \qquad (11.10)$$

and we may set the change in inputs ΔX proportional to the level of profits as long as $\Delta CVP > 0$:

$$\Delta X = IF\ \Delta CVP > 0\ THEN\ CVP * .01\ ELSE\ 0. \qquad (11.11)$$

Essentially the same signal can be formed by $\Delta CVP/DT = DERIVN(CVP, 1, .1)$.

To complete the STELLA model, we set up the module for the adjustments of input quantities as shown in Fig. 11.2.

Figure 11.3 shows the rise in profit as your company increases its use of wood in the production of toys. Figure 11.4 confirms our analytically derived condition (11.9) for the profit maximum; the cost of an additional unit of inputs increases as you expand the size of your operation and ultimately equals the price of a unit of output, as your company achieves the profit maximum. The maximum profit for your competitive firm is not zero ($1,562) when its optimal output (630 units) is reached. Additional producers of wooden toys may enter the market because a

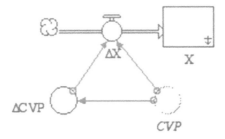

Fig. 11.2 Adjustments in input quantity

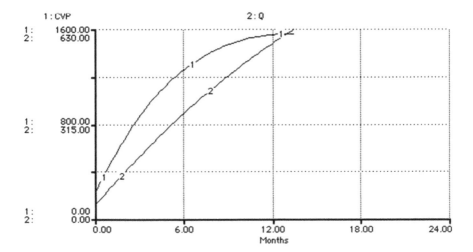

Fig. 11.3 Current value of profits and output

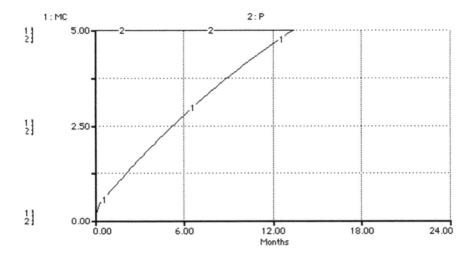

Fig. 11.4 Marginal cost and price

profit is to be made from selling wooden toys. As a consequence, they will drive down the price of the toys and collectively increase output. The number of toy producers is expected to increase to the point where the entry of an additional producer into the market leads to a profit rate of 0. In Chap. 14 we will explicitly solve for the equilibrium number of firms in a competitive market in which all firms produce an identical output but some may use different technology (*A* and ALPHA are not the same for each firm).

In our preceding model, we arbitrarily chose the rate of expansion of the company and set it proportional to the current value of profits. Choose alternative proportionality factors and observe the model results. Then, explicitly introduce labor as one of the essential inputs into the production of wooden toys. Of course, labor is costly, and you need to specify a wage rate at which you compensate the workers in your company. Analogous to the price of wood, assume that you have no influence on this wage rate and that it is an exogenous parameter in your model. Assume that the workers that you just hired need to be trained before they can contribute to the production process, and introduce an appropriate lag in the production function. Similarly, make the rate of expansion of inputs *X* a delayed function of the labor inputs. How may the use of tools and other capital equipment affect optimal output, the profit maximum, and your company's ability to expand its production?

In this chapter, we dealt with the ideal case of perfect competition and calculated the rate of expansion and optimum size of a firm in such a market. In the following chapter, we identify the optimum output level of the same firm if it can function as a monopoly. Both cases are extreme ones when compared with reality, but they enable us to delineate the realm of likely choices by firms. They are useful to make general statements about the desirability of alternative taxes (Chap. 12) or alternative market forms for the supply of essential, nonrenewable resources (Chaps. 21–23).

11.2   Competitive Firm Model Equations

```
X(t) = X(t - dt) + (ΔX) * dt
INIT X =  1  {Units of X}
INFLOWS:
ΔX =  IF ΔCVP > 0 THEN CVP*.01 ELSE 0   {Units of X per Time Period}
A =  50  {Units of Q per Units of X}
ALPHA = .5
C = W*X   {Dollars}
CVP = P*Q - C   {Dollars}
MC = (C - DELAY(C,DT,.1))/(Q - DELAY(Q,DT,.1))  {Dollars per Units of Q}
MR = (R - DELAY(R,DT,.1))/(Q - DELAY(Q,DT,.1))  {Dollars per Units of Q}
P = 5 {Dollars per Unit}
Q = A*X^ALPHA     {Units of Q}
R = P*Q   {Dollars}
W = 10
ΔCVP =  CVP - DELAY(CVP,DT,.1)   {Dollars}
```

Chapter 12
The Profit-Maximizing Monopoly

If two people agree all the time, one of them is unnecessary

David Mahoney

12.1 Introduction

In contrast to the previous chapter, let us assume that you are the proud owner of a firm that holds a monopoly on wooden toys. You are now in a position to simultaneously decide about the levels of output *and* sales price. However, your firm still competes for the input material wood on a competitive market. The price of wood used to produce toys is given and fixed. A second constraint imposed on you is given by the fact that even though you can set the sales price of your output, consumers will respond to a price change by adjusting the quantity of wooden toys they wish to buy. The relationship between the price that you set for your output and the quantity consumers are willing to buy can be specified with a demand curve. Typically, a demand curve is defined as a function that relates the price consumers are willing to pay to the quantity of a good available to them:

$$P = P(Q) \text{ with } \frac{\partial P}{\partial Q} < 0 \tag{12.1}$$

Given such an inverse relationship between price and demand, how many wooden toys should your firm produce and at what price should they be sold to maximize profits?

Before we attend to this question with a STELLA model, let us first derive the analytical solution. As in the previous chapter, we calculate the first partial

A save-disabled version of STELLA® and the computer models of this book are available at www.iseesystems.com/modelingeconomicsystems.

M. Ruth and B. Hannon, *Modeling Dynamic Economic Systems*,
Modeling Dynamic Systems, DOI 10.1007/978-1-4614-2209-9_12,
© Springer Science+Business Media, LLC 2012

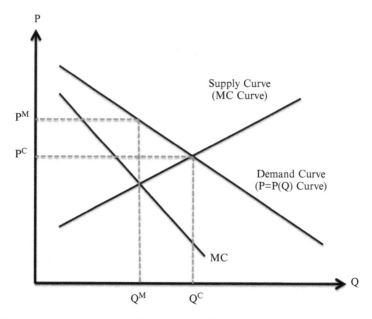

Fig. 12.1 Optimal output in competitive and monopolistic markets

derivative of the profit function with respect to a change in price. The profit function is now

$$\text{CVP} = R - C = P[Q(X)]^*Q(X) - W^*Q(X), \qquad (12.2)$$

which leads to

$$\frac{\partial \text{CVP}}{\partial Q} = \frac{\partial R}{\partial Q} - \frac{\partial C}{\partial Q} = \text{MR} - \text{MC} = 0 \Rightarrow \text{MR} = \text{MC}. \qquad (12.3)$$

Note that, since price is a function of Q, $\text{MR} \neq P$ in the case of the monopoly. Thus, in the profit maximum $\text{MR} = \text{MC} \neq P$. The difference between the marginal revenues and price in the profit maximum is known as the *monopoly rent rate*. This rate is shown for linear demand and supply curves in the following figure. As a monopolist, you would identify the profit-maximizing output level Q^M by setting $\text{MR} = \text{MC}$ and then charge the price P^M. The large set of firms in the perfectly competitive market, in contrast, would each set $P = \text{MC}$ resulting in Q^C as their optimal output, which they would sell at the price P^C.

Analogously to the previous chapter, we assume that the monopoly's expansion rate is determined by the level of the profit rate (Fig. 12.1).

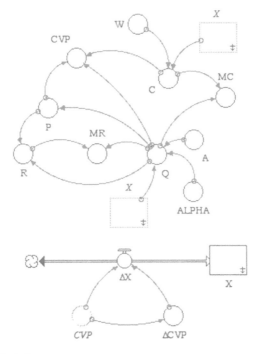

Fig. 12.2 Monopolistic firm

$$\Delta CVP = \text{IF } \Delta CVP > 0 \text{ THEN } CVP^*.01 \text{ ELSE } 0. \qquad (12.4)$$

The STELLA diagram is shown in Fig. 12.2. The results of Fig. 12.3 indicate that the optimal size of your monopolistic firm is reached in less than 6 months. The maximum profit is found to be $2,092 for an output of 432 units, at a price of $6.57 per unit output, when the marginal revenue and cost are $3.24. The graph of Fig. 12.4 confirms our analytical solution stated in (12.3) and shows a difference between price and marginal cost—monopoly rent rate—of $3.33.

Expand this model to include labor and capital in the production process and make the rate of expansion of your monopoly dependent on the use of labor and capital. Introduce a time lag at which your company can expand in response to a higher capital stock, and investigate the impacts of such lags on the change in the monopoly rent rate. Similarly, assess the implications of alternative technologies on maximum profit, optimal output, and monopoly rent rate. For example, change the coefficient ALPHA in the production function that we used

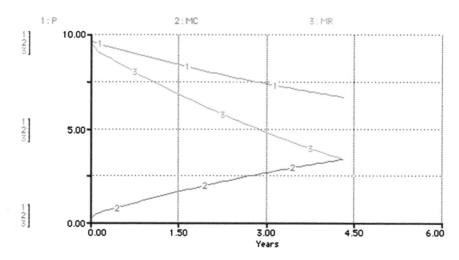

Fig. 12.4   Monopoly price, marginal cost, and marginal revenue

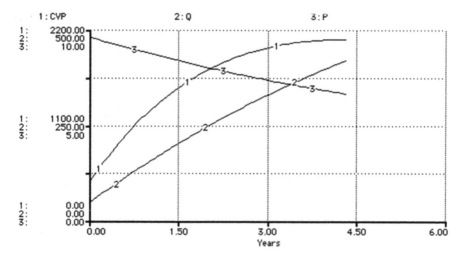

Fig. 12.3   Reaching the profit-maximizing conditions

earlier. Here, we have chosen ALPHA = 0.5. For 0 < ALPHA < 1, increases in inputs lead to an increase of outputs at a decreasing rate. This case is known as *decreasing returns to scale*. What will happen if ALPHA becomes greater than 1?

12.2   Monopoly Model Equations

X(t) = X(t - dt) + (ΔX) * dt
INIT X = 1 {Units of X}
INFLOWS:
ΔX = IF ΔCVP > 0 THEN CVP*.01 ELSE 0 {Units of X per
Time Period}
A = 50 {Units of Q per Units of X}
ALPHA = .5
C = W*X {$}
CVP = (P*Q-C) {$}
MC = (C - DELAY(C,DT,.01))/(Q - DELAY(Q,DT,.1)) {$ per
Units of Q}
MR = (R - DELAY(R,DT,10))/(Q - DELAY(Q,DT,.1))   {$ per
Units of Q}
P = 10-.00794*Q {The selling price is a function of Q; that is,
the monopoly controls the output level AND price, unlike the
competitive firm who could not affect price. This is the demand
curve, experimentally derived. $ per Unit Q}
Q = A*X^ALPHA {Units of Q}
R = P*Q {$}
W = 10 {$ per Units X}
ΔCVP = (CVP - DELAY(CVP,DT,1))

12.3   Effects of Taxes on Monopolistic Output and Price

In the preceding model, we have seen that the monopoly can capitalize on its market
power by charging a rent in excess of its marginal cost. When compared to the case
of perfect competition modeled in the previous chapter, we also see that the
monopoly restricts output below that of the competitive firm and sells that output
at a higher price. Thus, governments frequently intervene to affect a monopolist's
decisions on prices and quantities. Among the most preferred methods of intervention are taxes on output, profits, or as a lump sum. Each of these taxing schemes has
different affects on the monopolist's optimal choice of outputs and prices and thus
may achieve a government's goal of extending supply or reducing price to a
different extent. All of these taxing schemes, however, generate revenues, and we
will assess the impact of taxes on the monopolist's behavior under the assumption
of equal tax revenues for the government.

The STELLA model of Fig. 12.5 is based on the one used previously in the
absence of taxes. We modified the model to accompany the impacts of a PRODUC-
TION TAX of $0.8 per unit of Q and calculate the corresponding tax revenues in the
optimum of the firm. Similarly, we introduced a PROFIT TAX and a LUMP SUM
TAX whose values we set such that they generated the same tax revenue as the
profit tax of $0.8 per unit Q. Because the production tax has a direct influence on the
marginal revenues, we need to set up the model to calculate the new marginal
revenue NMR in case the government levies a production tax.

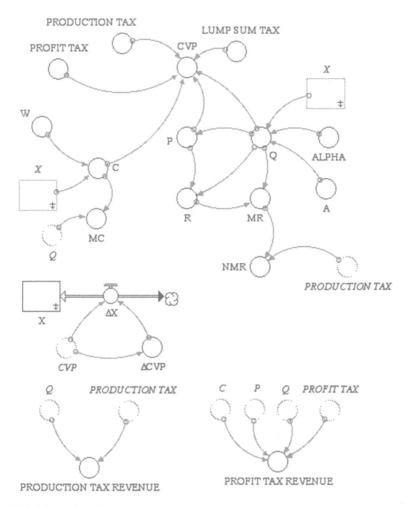

Fig. 12.5 Monopoly with taxes

The calculations of tax revenues in our STELLA model are not done with stocks, but with converters. If we use stocks, we calculate the cumulative tax revenue collected by the government over time. The size of this stock, of course, depends on the speed at which the monopolist adjusts production to achieve a profit maximum. The faster the profit maximizing output can be achieved, everything else being equal, the lower are the cumulative tax revenues. To calculate the tax revenues over time, as the monopolist gradually finds, the profit-maximizing output and price levels make use of stocks and flows, rather than transforming variables.

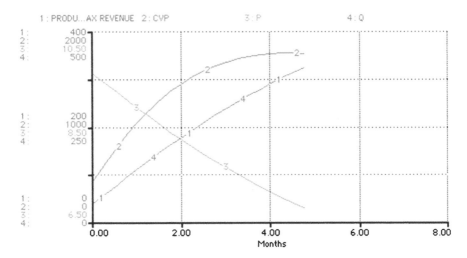

Fig. 12.6 Monopoly with production tax

Run the model in the absence of a tax at a DT $= 0.25$ and you will find that the maximum profits are CVP $= 2{,}093$, the profit maximizing-output level is $Q = 423$, which is sold on the market at a price $P = \$6.44$ per unit.

Now, introduce the production tax of $0.8 per unit. Figure 12.6 shows the effects of the production tax of $0.8 on profit, price, output quantity, and marginal cost. Tax revenues are approximately $314 (313.92, to be precise), profits are roughly $1,772, and the profit-maximizing output level and price are 392 and $6.88, respectively. Profits are now lower, prices higher, and optimum production also declined as compared to the non-tax case. Thus, the tax neither extended production nor decreased price for consumers. It did, of course, lower the monopolist's profits.

Figure 12.7 shows the results for a lump sum tax. Here, given our prerequisites for such a tax, government revenues are $313.92. The corresponding equilibrium price is $6.72. Output and profits are, respectively, 413 units and $1,779.

Next, introduce a profit tax that generates—at the monopolist's profit maximum—the same tax revenues as the production tax. You will find that that profit tax is $0.149945 per dollar profits earned by the monopolist to generate exactly the tax revenue of $313.92 as the production tax case did. The corresponding profits are $1,779, price is $6.64, and output is a little over 423. Again, profits and optimal output are lower than in the absence of taxes, price is higher.

Compare the output in Figs. 12.7 and 12.8. Note how the firm differently treats the Lump Sum vs. the Profit Tax. The first is a fixed reduction in profits while the second depends on the profit itself.

Calculate the cumulative tax revenues that are generated by the end of a year in the case of a production tax of $1.2 per unit output and then find the profit and lump sum taxes that generate the same cumulative tax revenues. Which of the three taxes lead to the smallest increase in price and which to the smallest decline in output

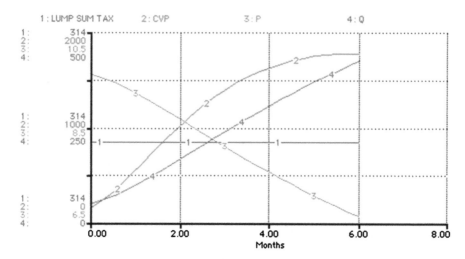

Fig. 12.7 Monopoly with lump sum tax

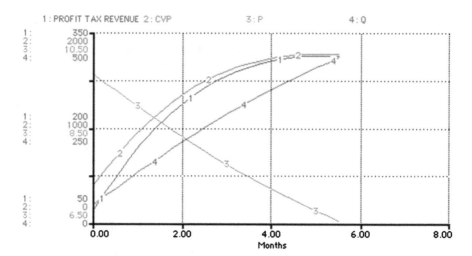

Fig. 12.8 Monopoly with profit tax

when compared to the no-tax case? Change the speed at which the monopolistic firm can adjust the size of its operation. How does a quicker adjustment affect your answer?

Now that we have seen the impact of a monopoly on output quantities and prices and the limited ability of the government to affect monopolistic behavior through taxes, let us investigate how the monopoly fares if its production results in pollution that has negative health effects on consumers. The following section addresses this issue.

12.4   Monopoly with Taxes Model Equations

X(t) = X(t - dt) + (ΔX) * dt
INIT X = 1 {Units of X}
INFLOWS:
ΔX = IF ΔCVP> 0 THEN CVP*.01 ELSE 0 {Units of X per
Time Period}
A = 50 {Units of Q per Units of X}
ALPHA = .5
C = W*X {Dollars per Time Period}
CVP = ((P-PRODUCTION_TAX)*Q-C)*(1-PROFIT_TAX) -
LUMP_SUM_TAX {Dollars per Time Period}
LUMP_SUM_TAX = 0 {$313.92 Dollars per Time Period}
MC = (C - DELAY(C,DT,.01))/(Q - DELAY(Q,DT,.1))   {Dol-
lars per Units of Q}
MR = (R - DELAY(R,DT,10))/(Q - DELAY(Q,DT,.1))   {Dollars
per Units of Q}
NMR = MR-PRODUCTION_TAX {Dollars per Units Q}
P = 10-.00794*Q {Dollars per Unit Q}
PRODUCTION_TAX = 0 {.8;   Dollars per Unit Q}
PRODUCTION_TAX_REVENUE = PRODUCTION_TAX*Q
{Dollars per Time Period}
PROFIT_TAX = 0 {0.149945 Dollars per Dollar Profit}
PROFIT_TAX_REVENUE = (P*Q-C)*PROFIT_TAX {Dollars
per Time Period}
Q = A*X^ALPHA {Units of Q}
R = P*Q {Dollars per Time Period}
W = 10 {Dollars per Unit X}
ΔCVP = (CVP - DELAY(CVP,DT,1))

12.5   Monopolistic Production and Pollution

Let us return to the basic model of a monopoly without tax and introduce a pollutant that is generated as a by-product from the production process and may need abatement. Whether or not abatement actually takes place, and the degree to which it takes place, depends on whether there is an incentive for the producer to incur the cost of abatement. Assume that the level of pollution depends on the output quantity Q of the desired product and the extent to which abatement takes place. The relationship is given by the graph of Fig. 12.9.

The extent of abatement efforts, in turn, is dependent on the monopolist's choice of a CONTROL FACTOR (Fig. 12.10). If this factor is 0, pollution is emitted without control or control costs. The control factor is not a fraction because it can exceed 1.

The level of pollution is assumed to affect the health of consumers. Zero pollution levels result in zero adverse health effects. But zero pollution can be achieved only if no production takes place. Low, nonzero pollution levels require

Fig. 12.9 Pollution–
abatement relationship

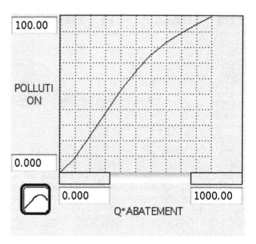

Fig. 12.10 Control factor
for pollution abatement

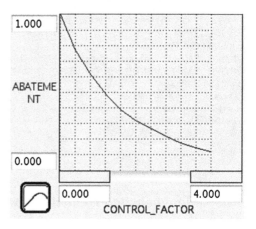

high abatement efforts, that is, a choice of a high CONTROL FACTOR. If the
monopolist chooses to significantly control pollution, then pollution cost is high:

$$POLLUTION\ COST\ =\ CONTROL\ FACTOR * Q \qquad (12.5)$$

We have assumed here that costs of abatement are directly proportional to
production. No doubt this is not true, but any more exact connection would not
change the message of this model. Note that this cost must be included in the profit
calculation and the calculation of the total and marginal costs.

If the monopolist chooses a low control factor, pollution levels and adverse
health effects will be high (Fig. 12.11). As a consequence, consumers may be
willing to pay less for, or buy less of, the monopolist's product. The effect of health
cost on the producer may occur in the following way: Increased expenditures on
health care due to increased pollution mean that the consumers will have less

Fig. 12.11 Health effects
of pollution

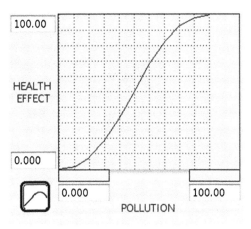

Fig. 12.12 Demand effects
from changes in the health
effect of pollution

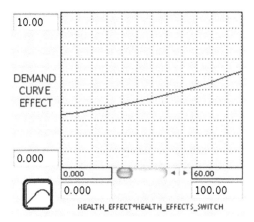

money to spend on the producer's product. This effect is arranged through a change
in the slope of the demand curve for the product—as pollution increases, the slope
of the demand curve increases (the choke-off price stays the same). This means that
increased pollution leads to increased health expenditures that, in turn, lead to
decreased expenditures on the product. The DEMAND CURVE EFFECT that
results from the HEALTH EFFECT of pollution is specified graphically in our
model (Fig. 12.12). Under these assumptions, it may behoove the producer to spend
on pollution control, to maximize its profits, given this pernicious effect!

We have introduced a HEALTH EFFECTS SWITCH in the model (Fig. 12.13)
that enables you to easily control the impacts of health effects on the demand curve.
If the switch is 1, the health effects are lowering the consumer's demand curve. If it
is 0 (and the control factor is 0), we should have the same effect as we had in the
basic model of the monopoly, where we did not even consider the pollution or
environmental problems. That condition is the "no pollution" result, and it forms a
kind of benchmark for the profit level of this company.

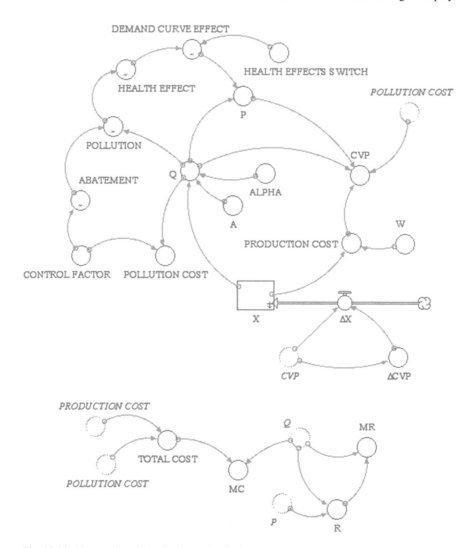

Fig. 12.13 Monopolist with pollution and pollution cost

Run the model for a DT = 0.125, the following specification for the change in inputs

$$\Delta X = IF \ \Delta CVP > 0 \ THEN \ .008^*CVP \ ELSE \ 0 \qquad (12.6)$$

and different values of the CONTROL FACTOR. Observe the resulting CVP, Q, and POLLUTION. No pollution control leads to CVP = 2,443 and a control factor of 2 results in CVP = 2,116. You will find that the highest CVP can be achieved if there is moderate pollution control. A pollution control factor of 0.404 will yield the highest CVP. The graph in Fig. 12.14 shows the optimal case.

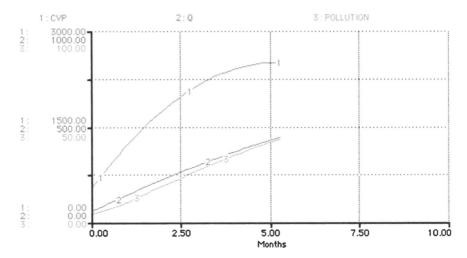

Fig. 12.14 Profit maximization for the monopolist with pollution

At a given CONTROL FACTOR the greatest possible profit may not be the overall maximum. Here, then, we have the producer adjusting its pollution control factor to reduce health effects of unabated pollution so that the consumers will divert some of these former expenditures on health over to buying its product. It is simply the monopolist's goal of maximizing profits—not altruism—that leads it to cut down on pollution. The acknowledgment of this position seems too inhumane and too calculating, but any good policy maker, private or public, needs to know of the possibility of the existence of such an operating point.

It is easy for the monopolist to act in self-interest here and realize that reducing pollution somewhat will lead to an increase in profits. But what about the competitive market? There, no one producer can act without loss of market share and profits. Because one cannot act alone, none will act. This case calls out for governmental control.

We have now seen that monopolistic firms will reduce output and sell at higher prices than competitive firms, but this is not necessarily "bad" for consumers. If production leads to the generation of waste products, it may be even preferred to have a profit-maximizing monopolist who has significant power to adjust output quantities and control pollution. To make a judgment on the desirability of such a monopoly, however, requires that we know how consumers value the goods produced by the monopoly vs. pollution.

Of course, other factors need to be considered before we take sides in favor or against monopolistic behavior. One of these is employment in the monopolistic firm. If the monopoly is large and operates multiple facilities, it has considerable power over the distribution of output and employment among its various facilities. As a consequence, the monopoly has significant influence on the development and decline of the regions in which its factories are located. We model the case of a monopolist with two production facilities in the following chapter.

12.6   Monopolistic Production and Pollution Model Equations

X(t) = X(t - dt) + (ΔX) * dt
INIT X = 1.5 {Units of X}
INFLOWS:
ΔX = IF ΔCVP > 0 THEN .008*CVP ELSE 0 {Units of X per
Time Period}
A = 50 {Units of Q per Units of X}
ABATEMENT = GRAPH(CONTROL_FACTOR)
(0.00, 1.00), (0.4, 0.77), (0.8, 0.615), (1.20, 0.49), (1.60, 0.39),
(2.00, 0.32), (2.40, 0.27), (2.80, 0.22), (3.20, 0.175), (3.60,
0.145), (4.00, 0.12)
ALPHA = .5
CONTROL_FACTOR = 2 {.404}
CVP = (P*Q-PRODUCTION_COST-POLLUTION_COST) {$}
DEMAND_CURVE_EFFECT =
GRAPH(HEALTH_EFFECT*HEALTH_EFFECTS_SWITCH)
(0.00, 3.38), (5.00, 3.50), (10.0, 3.70), (15.0, 3.85), (20.0,
4.05), (25.0, 4.25), (30.0, 4.45), (35.0, 4.70), (40.0, 4.95),
(45.0, 5.20), (50.0, 5.50), (55.0, 5.85), (60.0, 6.20), (65.0,
6.55), (70.0, 6.95), (75.0, 7.35), (80.0, 7.80), (85.0, 8.30),
(90.0, 8.85), (95.0, 9.40), (100, 10.0)
HEALTH_EFFECT = GRAPH(POLLUTION)
(0.00, 0.00), (10.0, 1.50), (20.0, 7.50), (30.0, 18.5), (40.0,
33.0), (50.0, 50.0), (60.0, 68.0), (70.0, 82.0), (80.0, 93.0),
(90.0, 98.0), (100, 100)
HEALTH_EFFECTS_SWITCH = 1
MC = (TOTAL_COST - DELAY(TOTAL_COST,DT,.01))/(Q -
DELAY(Q,DT,.1)) {$ per Units of Q}
MR = (R - DELAY(R,DT,10))/(Q - DELAY(Q,DT,.1))   {$ per
Units of Q}
P = 10-DEMAND_CURVE_EFFECT/1000*Q {$ per Unit Q}
POLLUTION = GRAPH(Q*ABATEMENT)
(0.00, 0.00), (100, 9.00), (200, 23.0), (300, 37.5), (400, 52.0),
(500, 64.5), (600, 75.5), (700, 83.5), (800, 89.5), (900, 95.0),
(1000, 100)
POLLUTION_COST = CONTROL_FACTOR*Q
PRODUCTION_COST = W*X {$}

Q = A*X^ALPHA {Units of Q}
R = P*Q {$}
TOTAL_COST = POLLUTION_COST+PRODUCTION_COST
W = 10 {$ per Units X}
ΔCVP = CVP - DELAY(CVP,DT,1)

Chapter 13
Monopolistic Collusion

The book of Nature is the book of Fate. She turns the gigantic pages—leaf after leaf, never returning one.

R. W. Emerson, Everyman's Library Edition, 1915, p. 157.

13.1   Joint Management of Two Monopolists

Assume two production plants are jointly managed. The decision to jointly manage may arise from at least the following two scenarios. First, a monopoly might own two different kinds of plants, such as electric generating stations; one is old and operates with high cost; the other is new and operates at low cost. The model developed here can tell the monopolist how to divide output between the plants, assuming that both facilities can produce the desired output. On the darker side, we may imagine this setting arises as the result of two former competitors deciding to collude and treat their market as though they were a monopoly. The model can tell them how to allocate production to maximize their joint profits.

Output of the firms is $Q1$ and $Q2$, respectively. Together, the two plants control the entire market for their output $Q = Q1 + Q2$. Production functions and costs are different for the two plants. Production functions are specified as

$$Q1 = A1^*X1^{\wedge}\text{ALPHA1} \tag{13.1}$$

and

$$Q2 = A2^*X2^{\wedge}\text{ALPHA2}, \tag{13.2}$$

A save-disabled version of STELLA® and the computer models of this book are available at www.iseesystems.com/modelingeconomicsystems.

M. Ruth and B. Hannon, *Modeling Dynamic Economic Systems*,
Modeling Dynamic Systems, DOI 10.1007/978-1-4614-2209-9_13,
© Springer Science+Business Media, LLC 2012

with the constants $A1 = 50, A2 = 60$, ALPHA1 $= 0.6$, and ALPHA2 $= 0.5$. Input prices W are the same for both firms. Thus, costs depend on only the input quantities $X1$ and $X2$:

$$C1 = W^*X1 \tag{13.3}$$

$$C2 = W^*X2. \tag{13.4}$$

The combined current value of profits from both firms is

$$\begin{aligned} CVP &= CVP1 + CVP2 = P^*(Q1 + Q2) - C1 - C2 \\ &= R1 + R2 - C1 - C2 \end{aligned} \tag{13.5}$$

and the profit maximizing output levels are defined by

$$\frac{\partial CVP}{\partial Q_1} = MR1 - MC1 = 0, \tag{13.6}$$

$$\frac{\partial CVP}{\partial Q_2} = MR2 - MC2 = 0 \Rightarrow MR2 = MC2 \tag{13.7}$$

and we must have

$$MR1 = MR2. \tag{13.8}$$

In our model, the demand curve for the product is linear and because $Q = Q1 + Q2$. However, we set up the marginal revenue specifications such that we can easily accommodate nonlinear demand functions. The results that follow are shown for the simple case of

$$P = 10 - 0.00794^*(Q1 + Q2). \tag{13.9}$$

Changes in input quantities $X1$ and $X2$ to achieve a profit-maximizing output level may now be defined on the basis of the difference between marginal revenues and marginal cost. Once marginal revenues equal marginal cost, no change in input quantities should occur because a maximum for the profits is encountered. In this model, we choose graphical functions to change input quantities, allowing $X1$ and $X2$ to overshoot and return to their equilibrium values. The graphical function for the change in $X1$, $\Delta X1$ is defined as shown in Fig. 13.1 and the change in $X2$, $\Delta X2$ is specified analogously.

Of course, for real firms the actual paths for approaching equilibrium are determined, for example, by the availability of the necessary capital, construction progress, and the time it takes to phase in new capital. It is possible to represent these influences with $\Delta X1$ and $\Delta X2$, as shown in Fig. 13.2.

The model shows profit-maximizing output levels of $Q1 = 220.6$ (Fig. 13.3) and $Q2 = 1616.0$ (Fig. 13.4) with input quantities of $X1 = 11.9$ and $X2 = 7.3$. The maximum profit level is approximately \$2,471 per month (Fig. 13.5). Note how one of the profit functions overshoots and backs down, even though the maximum

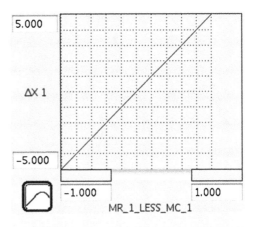

Fig. 13.1 Changes in input quantity in production Plant 1

profit rises continuously. The overshoot is really undesirable because it means idling part of the plant that had been built. This idling may not have actually happened, however, if the overshoot were not too long. No plant operates at 100% capacity forever, but it may be possible to operate higher than normal for a short time; for example, by having workers work overtime. Can you adjust the problem to eliminate the overshoot?

Let us return to the interpretation in which the setting of the model is the collusion of two former competitors joining to form a monopoly. The solution of the model tells the producers their profit-maximizing output levels and the distribution of production and profits between the two production plants. The model also shows that CVP1 is greater than CVP2. Furthermore, the combined profits under a monopolistic collusion are higher than profits of perfect competitors. The difference in profits between firms may be a reason for conflict between the two plants, and the arrangements between the two producers may break down unless they can lay out some suitable way to split the greater profits. Can you define a set of profit division schemes that would or should hold the two together?

Let you think we encourage you too much toward law-breaking plans, realize that you have to know them to stop them. Lest you think that we direct you only into jobs with the Justice Department, realize that you would make a better business consultant if you could solve the problems laid out here.

The model may also get handy if you end up working for Japanese industry. Nippon is the largest steel maker in the world. Nippon has many plants of many different ages and thus of different costs of operation. When world demand for steel is high, all plants proceed at full speed. But as demand falls, the managers consider two options: the first is to shut down the high-cost plants until the demand can be met by the low-cost plants, producing the steel for the lowest possible cost. Shutting down plants, however, requires layoffs and leads to increasing government welfare and unemployment costs. Or, they could slow down production of each

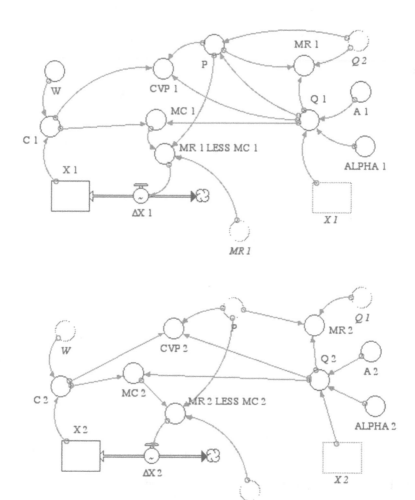

Fig. 13.2 Monopolistic collusion

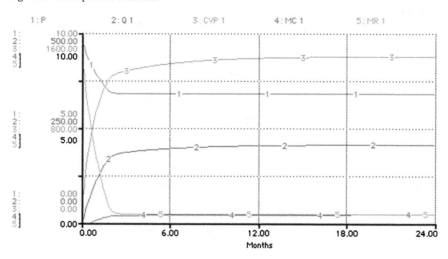

Fig. 13.3 Finding optimal conditions for production Plant 1

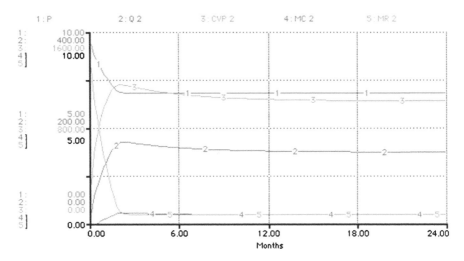

Fig. 13.4 Finding optimal conditions for production Plant 2

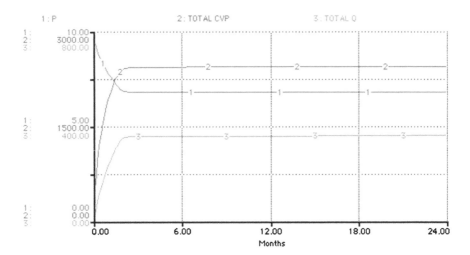

Fig. 13.5 Price, profit, and output with collusion

plant at the same percentage until demand is met. This solution keeps everyone employed but with lower bonus payments and higher operating costs, and more costly steel.

The two options may be presented to the Japanese government, which is faced with the choice between incurring increased welfare payments or covering the difference in cost to exercise the second option. The government has frequently chosen the second option. When it does, US steelmakers complain about the Japanese government subsidizing exports of steel, seemingly in support of monopolistic collusion! They neglect to see that when US steelmakers close down their high-cost plants, the US government picks up the welfare and unemployment cost, and the unemployed pay their own relocation costs, both of which amount to a subsidy to US steel! Additionally, valuable skills may be lost when US steel plants shut down and workers are laid off. By the time orders for US steel increase, those high-skilled workers may have left the work force or be employed elsewhere in the economy.

13.2   Monopolistic Collusion Model Equations

```
X_1(t) = X_1(t - dt) + (ΔX_1) * dt
INIT X_1 = .5   {Units of Input}
INFLOWS:
ΔX_1 = GRAPH(MR_1_LESS_MC_1)
(-1.00, -5.00), (-0.8, -4.00), (-0.6, -3.00), (-0.4, -2.00), (-0.2, -1.00), (-5.55e-17,
0.00), (0.2, 1.00), (0.4, 2.00), (0.6, 3.00), (0.8, 4.00), (1.00, 5.00)
X_2(t) = X_2(t - dt) + (ΔX_2) * dt
INIT X_2 = .5   {Units of Input}
INFLOWS:
ΔX_2 = GRAPH(MR_2_LESS_MC_2)
(-1.00, -5.00), (-0.8, -4.00), (-0.6, -3.00), (-0.4, -2.00), (-0.2, -1.00), (-6.66e-17,
0.00), (0.2, 1.00), (0.4, 2.00), (0.6, 3.00), (0.8, 4.00), (1, 5.00)
ALPHA_1 = .6

ALPHA_2 = .5
A_1 = 50 {Units of output per Unit of Input}
A_2 = 60 {Units of output per Unit of Input}
CVP_1 = P*Q_1-C_1 {Dollars per Time Period}
CVP_2 = P*Q_2-C_2 {Dollars per Time Period}
C_1 = W*X_1 {Dollars per Time Period}
C_2 = W*X_2 {Dollars per Time Period}
MC_1 = (C_1-DELAY(C_1,DT,.1))/(Q_1-DELAY(Q_1,DT,.1))  {Dollars per Unit
Output}
MC_2 = (C_2-DELAY(C_2,DT,.1))/(Q_2-DELAY(Q_2,DT,.1))  {Dollars per Unit
Output}
MR_1 = P-.01588*(Q_1 + Q_2)
MR_1_LESS_MC_1 = MR_1-MC_1 +0*P  {substitue P for MR here and in the
other driver for competitive solution.}
MR_2 = P-.01588*(Q_1+Q_2)     {Dollars per Unit Output}
MR_2_LESS_MC_2 = MR_2-MC_2 +0*P{substitue P for MR here and in the
other driver for competitive solution.}
P = 10 - .00794*(Q_1 + Q_2) {Dollars per Unit Output}
Q_1 = A_1*X_1^ALPHA_1 {Units of Output per Time Period}
Q_2 = A_2*X_2^ALPHA_2 {Units of Output per Time Period}
TOTAL_CVP = CVP_1+CVP_2 {Dollars per Time Period}
TOTAL_Q = Q_1+Q_2 {Units of Output per Time Period}
W = 10 {Dollars per Unit Input}
```

Chapter 14
Quasi-Competitive Equilibrium

> *There is no likelihood man can ever tap the power of the atom.*
>
> Robert Millikan, Nobel Laureate, Physics, 1923

14.1 Finding the Number of Profit-Maximizing Competitors

In previous chapters we have considered two extreme market forms—perfect competition and monopolistic behavior. Many markets, however, fall between these two extremes. Some companies are large and well endowed and therefore have significant market power. Yet, there are few of those fortunate ones. A larger group of less profitable firms compete for the part of the market not served by the large firms. As a result of these differences among firms, we do not have perfect competition. Rather, only a limited number of firms can compete successfully. In this chapter, we determine how many profit-maximizing firms can be successful in a quasi-competitive market. In our model, we assume that there are two groups of firms in that market. The first group consists of $N1$ firms. Each of those firms has the same production technology to produce $Q1$ units of the product. Their production function is

$$Q1 = A1 * X1^2 - \text{ALPHA1} * X1^3, \tag{14.1}$$

with $A1 = 50$ and $\text{ALPHA1} = 13$ as fixed parameters.

A save-disabled version of STELLA® and the computer models of this book are available at www.iseesystems.com/modelingeconomicsystems.

M. Ruth and B. Hannon, *Modeling Dynamic Economic Systems*,
Modeling Dynamic Systems, DOI 10.1007/978-1-4614-2209-9_14,
© Springer Science+Business Media, LLC 2012

In the second group of firms are exactly five firms. These firms are more efficient in producing the output quantity $Q2$ of the same good produced by firms in group 1. The production function of each of those firms is

$$Q2 = A2 * X2^2 - \text{ALPHA2} * X2^3, \qquad (14.2)$$

with $A2 = 75$ and $\text{ALPHA2} = 13$ as fixed parameters. The firms in this group are able to produce the product at lower cost. We can think of them as mining a rich vein of ore that is easy to reach. Those firms with $A = 50$ are mining the same type of ore, but it is not as easy to extract; their costs per ton are higher. When an equilibrium price for the ore is reached, how many of the miners with high extraction costs will be forced off the market? What will then be the profit for mines with a rich vein of ore?

Since firms in both groups produce the product, we need to sum their output to calculate the total quantity available to meet demand. The demand curve in our case is

$$P = 10 - 0.00338(N1 * Q1 + N2 * Q2) \qquad (14.3)$$

Fierce competition can force down a firm's production level until the market price of its products equals the average cost of manufacturing them. The firm's profit is reduced to zero. Although the firm may generate just enough revenues to cover its cost, from the perspective of economic theory, it is maximizing its profit. Unfortunately, the average cost, AC, of production, the marginal cost, MC, and the price, P, are all the same: profits are reduced to nothing.

The implications of the $P = \text{AC} = \text{MC}$ rule for market equilibrium are rather straightforward for our simple, quasi-competitive setting with two groups of firms. Because we know $N2 = 5$, we can easily calculate the number $N1$ of firms in group 1, in the market equilibrium, by solving the demand curve of (14.3) for $N1$. Then we need to recognize that the high-cost firms must follow the $P = \text{AC} = \text{MC}$ rule. Firms in group 1 compete heavily with each other. To determine their optimal output level, each of them has to set its marginal cost equal to the market price. Denoting MC1 as the marginal cost of the high cost firms and substituting MC1 for P in (14.3), we get from the demand curve,

$$N1 = \frac{10 - \text{MC1}}{0.00338 * Q1} - N2 * \frac{Q2}{Q1}. \qquad (14.4)$$

Firms in group 2 do not have to comply with this rule. Their number is fixed to $N2 = 5$, and they can realize a nonzero profit because they use more efficient technology or are endowed with higher quality resources.

To find the number of firms in group 1 we need to drive the difference between average cost and marginal cost to 0 ($\text{AC1} = \text{MC1}$) while insisting that each firm is maximizing profits ($P = \text{MC1}$). By increasing $X1$, we exhaust all profits through an increase in the number of firms in the market. Figure 14.1 shows that part of the STELLA model that is set up to calculate changes in $X1$ and thus $N1$.

The adjustments in production of firms in group 2 and the calculation of the demand curve and current value profits, CVP1, CVP2, are done in that part of

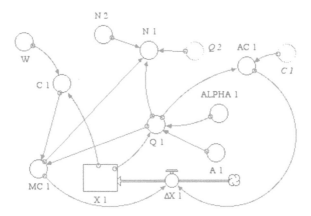

Fig. 14.1 Quasi-competitive equilibrium

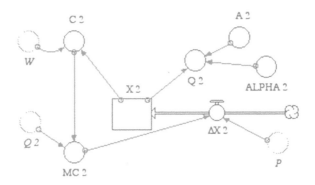

Fig. 14.2 Adjustments in production quantities

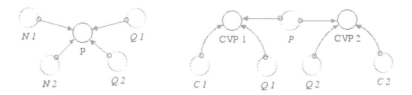

Fig. 14.3 Calculation of price and profits

the model shown in Fig. 14.2. Prices and profits are calculated in the module of Fig. 14.3.

Figure 14.4 shows that the optimal number of firms in group 2 is about nine. Figure 14.5 depicts the continuous equality of price and MC—maximum possible profits for each firm—and the eventual equality of MC and AC when the number of firms entering the market has stabilized. This is the equilibrium for the market. All high-cost firms reach a maximum of zero profits, while low-cost firms have positive profits (Fig. 14.6).

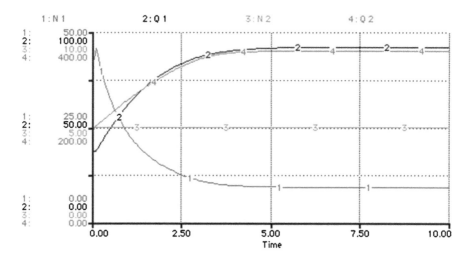

Fig. 14.4   The number of firms and their output

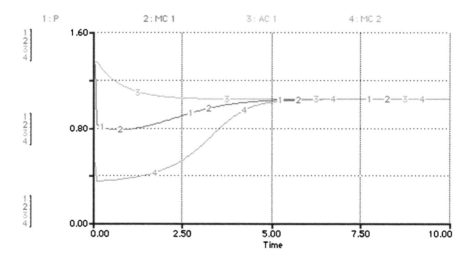

Fig. 14.5   Price, marginal costs, and average cost

Run the model at a DT $= 0.25$. Set up scatter plots for the production functions $Q1 = Q1(X1)$ and $Q2 = Q2(X2)$. Similarly, plot the cost functions in a scatter plot. Interpret the behavior of these functions and compare them for the two groups of mining companies.

We have now set up a model that is more realistic than either of the two extreme assumptions of perfect competition or a monopoly. Before we move on to relax another assumption frequently made in economic theory, we will explore this model a bit further. But rather than expanding on the model itself, let us perform sensitivity analyses on its parameters. We can do such sensitivity analyses with STELLA, as we have seen in Chap. 3.

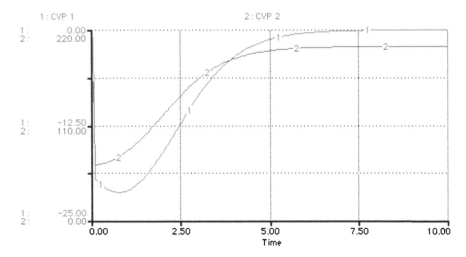

Fig. 14.6 Profits

14.2   Quasi-Competitive Equilibrium Model Equations

X_1(t) = X_1(t - dt) + (ΔX_1) * dt
INIT X_1 = 1 {Units of X}
INFLOWS:
ΔX_1 = AC_1 - MC_1 {Units of X per Time Period}
X_2(t) = X_2(t - dt) + (ΔX_2) * dt
INIT X_2 = 2 {Units of X}
INFLOWS:
ΔX_2 = P-MC_2 {Units of X per Time Period}
AC_1 = C_1/Q_1 {Dollars per Unit of Q}
ALPHA_1 = 13 {Units of Q per Unit X}
ALPHA_2 = 13 {Units of Output per Unit Input}
A_1 = 50 {Units of Q per Unit X}
A_2 = 75 {Units of Output per Unit Input}
CVP_1 = P*Q_1-C_1 {Dollars}
CVP_2 = P*Q_2-C_2 {Dollars}
C_1 = W*X_1 {Dollars}
C_2 = W*X_2 {Dollars}
MC_1 = (C_1-DELAY(C_1,DT,1))/(Q_1-
DELAY(Q_1,DT,.001)) {Dollars per Unit of Q}
MC_2 = (C_2-DELAY(C_2,DT,1))/(Q_2-
DELAY(Q_2,DT,.001)) {Dollars per Unit of Q}
N_1 = (10-MC_1)/(.00338*Q_1) - N_2*Q_2/Q_1   {Number of
High-Cost Firms}
N_2 = 5 {Number of Low Cost Firms}
P = 10 - .00338*(N_1*Q_1 + N_2*Q_2)

Q_1 = A_1*X_1^2 - ALPHA_1*X_1^3 {Units of Output}
Q_2 = A_2*X_2^2 - ALPHA_2*X_2^3 {Units of Output}
W = 50 {Dollars per Unit X}

Chapter 15
Modeling Economic Games

Heavier than air flying machines are impossible.

Lord Kelvin, Royal Society, c. 1895

15.1 Arms Race

At the outset of our discussion of the behavior of firms in Chap. 11, we stressed the assumption that the behavior of firms and consumers is coordinated in markets via price signals. Of course, supply and demand curves need not be as well behaved as we assumed them so far. There may not be a demand curve at all, for example; rather one may be evolving as exchange takes place in the marketplace.

As different parties meet in the market, a game unfolds in which each player may pursue a different strategy. As in the games you may have played just for fun, the "players" in an economic game often lack perfect information that would enable them to identify the best strategy. Specifically, they may not have perfect information on important characteristics of the other players.

In this section, we begin modeling games played by a small number of economic agents. The first of these games is the traditional arms race game (Hanneman 1988) that can easily be interpreted in an economic context. In fact, this mimetic behavior is analogous to the household consumer goods race.

In the following section, we model the exchange of goods in a simple barter economy. For simplicity, both the arms race game and the exchange economy are modeled for the case of two players. A more complex, multiplayer game is presented in the third section of this chapter, in which we model an auction game.

Let us start with a simple model of an arms race. The two competing nations are X and Y. Each strives to maintain 10% more arms than the other. This behavioral

---

A save-disabled version of STELLA® and the computer models of this book are available at www.iseesystems.com/modelingeconomicsystems.

M. Ruth and B. Hannon, *Modeling Dynamic Economic Systems*,
Modeling Dynamic Systems, DOI 10.1007/978-1-4614-2209-9_15,
© Springer Science+Business Media, LLC 2012

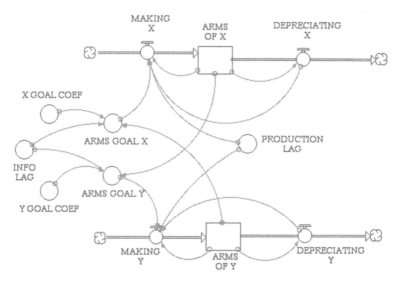

Fig. 15.1 Arms race model

assumption is captured in the STELLA model by the coefficients X GOAL COEFF and Y GOAL COEFF. We will change these coefficients to assess their impact on the outcome of the arms race.

Each country spies on the other to determine its level of arms, but there is a delay in the receipt of that information. The information gained by X about the level of Y's arms is delayed by three time steps. It is used to set the goal for arms production:

$$\text{ARMS GOAL } X = X \text{ GOAL COEF}^*\text{DELAY}(Y, 3), \qquad (15.1)$$

$$\text{ARMS GOAL } Y = Y \text{ GOAL COEF}^*\text{DELAY}(X, 3). \qquad (15.2)$$

When the spies do report in, each country begins to make arms. Each nation tries to close its perception of the arms gap by producing new arms and replacing those that depreciated. For the model, we assume that there is a manufacturing delay of three time steps:

$$\text{MAKE } X = \text{DELAY}(\text{MAX}(\text{ARMS GOAL } X - X + \text{DEPRECIATE } X, 0), 3),$$
$$(15.3)$$

$$\text{MAKE } Y = \text{DELAY}(\text{MAX}(\text{ARMS GOAL } Y - Y + \text{DEPRECIATE } Y, 0), 3).$$
$$(15.4)$$

The depreciation of arms is the same for each nation and set to 5% of its current stock. Equations (15.3) and (15.4) calculate the maximum value of the added arms or 0 to prevent the sale of arms outside the system, during a time of defusing of the race—the case in which the goal coefficients are less than 1 (Fig. 15.1).

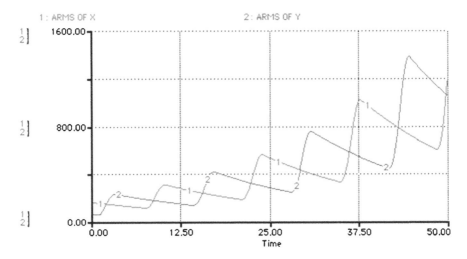

Fig. 15.2 Outcome of the arms race

Run the model with initial values of arms for $X = 150$ units and $Y = 50$ units. Choose a DT $= 0.125$. Figure 15.2 shows the results. Nation Y will initially increase its production of arms significantly to make up for the big difference in initial endowments, whereas X can afford to have some of its arms depreciate. However, soon X will try to catch up with Y by increasing its production. The result for the non-delayed case shows that stocks of arms for the two nations that fluctuate around each other and exponentially grow.

Alter the lags at which information is received by the two nations on the competing nation's stock of arms. What are the effects of these lags on the escalation of the race? Next, change the manufacturing lags and observe their impact on the arms race. You should find that the manufacturing delay is very important. For example, with no manufacturing delays and a difference in the set of initial arms, the arms levels oscillate but then converge and rise exponentially. As the lag time grows, the arms levels oscillate wildly, without growing divergence. With a lag of three time units, the two countries can even be trying to de-escalate the race and the arms levels still grow, indicating that even with good intentions, the race can continue. Confirm this finding by changing the X GOAL COEFF and Y GOAL COEFF to values larger than 0 but less than 1.

Try to introduce an arms treaty whose effect is to increase the retirement rate of arms. Try to imagine other ways of stopping the escalation of arms such as very low goal coefficients. Set the model to have one country be industrially more advanced than the other (e.g., differing lags in manufacturing between the two countries) and with better spies. Make interpretations of, and applications for, this kind of model in economics or business.

15.2   Arms Race Model Equations

X(t) = X(t - dt) + (MAKE_X - DEPRECIATE_X) * dt
INIT X = 150 {Arms}

INFLOWS:
MAKE_X = DELAY(MAX(ARMS_GOAL_X-
X+DEPRECIATE_X,0),0,3) {Arms per Time Period}
OUTFLOWS:
DEPRECIATE_X = .05*X {Arms per Time Period}
Y(t) = Y(t - dt) + (MAKE_Y - DEPRECIATE_Y) * dt
INIT Y = 50 {Arms}

INFLOWS:
MAKE_Y = DELAY(MAX(ARMS_GOAL_Y-
Y+DEPRECIATE_Y,0),0,3) {Arms per Time Period}
OUTFLOWS:
DEPRECIATE_Y = .05*Y {Arms per Time Period}
ARMS_GOAL_X = X_GOAL_COEF*DELAY(Y,3)
ARMS_GOAL_Y = Y_GOAL_COEF*DELAY(X,3)
X_GOAL_COEF = 1.1
Y_GOAL_COEF = 1.1

15.3   Barter Economy

So far we have focused in this book almost entirely on the market behavior of firms. Now let us model in more detail a game with a long history in economic theory, a "game" that we all played at one time or another: the exchange of goods or services that is not accompanied by an exchange of money. A sister helps you fix your bike and in exchange you help her with her homework. A neighbor watches your kids and in exchange you mow his lawn.

To model the barter economy, consider two people, A and B, who each possess an endowment of two goods X and Y. Person A possesses more of good X and person B possesses more of good Y. The preferences of each are given by the following utility functions:

$$UA = (XA^{\wedge}(1/3)^{*}YA^{\wedge}(2/3))/(XA + YA), \qquad (15.5)$$

$$UB = (XB^{\wedge}(2/3)^{*}YB^{\wedge}(1/3))/(XB + YB). \qquad (15.6)$$

With these utility functions and the initial endowments, the two players would like to have more of the good of which each possesses less. Therefore, we would expect them to exchange goods to increase their utility. For example, person A would exchange EX units of X if there was a positive change in utility ΔUA. If there was a loss in utility, then person A would ask EX units back from person B. We set

$$EX = 0.25 + .1^{*}\text{CHANGE } UA, \qquad (15.7)$$

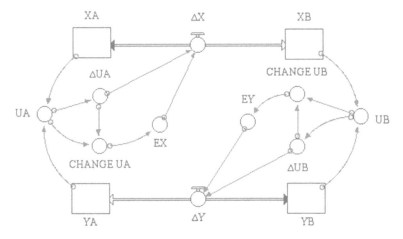

Fig. 15.3 Barter economy

with CHANGE *UA* as the percentage increase in player A's utility. If there was no change in utility, A should trade no further. An analogous relationship is assumed for player B.

The change ΔX in the players' endowments of X is

$$\Delta X = \text{IF } \Delta UA \geq 0 \text{ THEN } EX \text{ ELSE } - EX. \tag{15.8}$$

Because any player would give up some of one good only in exchange for some of the other good, we set

$$\Delta Y = \text{IF } \Delta UB \geq 0 \text{ THEN } EY \text{ ELSE } - -EY. \tag{15.9}$$

The resulting STELLA model is shown in Fig. 15.3.

Run the model with $XA = 30$, $YA = 15$ and $XB = 5$, $YB = 40$ at DT $= 0.25$. From the structure of the utility functions, it is clear that A prefers Y and B prefers X. We have set the initial values in opposition to these preferences to show how the model adjusts to the optimum distribution of trade (Fig. 15.4). Both of the players increase their utility through exchange (Fig. 15.5).

Can you find the equilibrium distribution of the two goods if both players have the same preference for those two goods? For example, assume

$$UA = (XA^{\wedge}(1/2)^{*}YA^{\wedge}(1/2))/(XA + YA), \tag{15.10}$$

$$UB = (XB^{\wedge}(1/2)^{*}YB^{\wedge}(1/2))/(XB + YB). \tag{15.11}$$

Think your answer through before running the changed model.

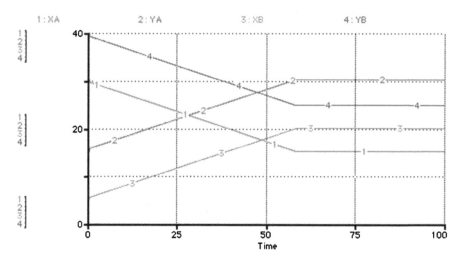

Fig. 15.4 Redistribution through exchange

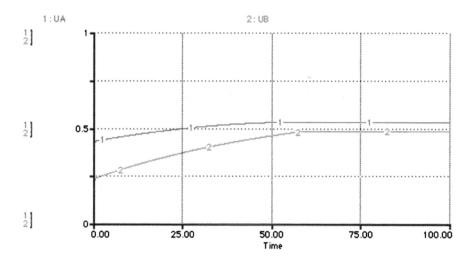

Fig. 15.5 Utility increase through exchange

Assume that person A has the utility function as in (15.5) but person B is envious of person A's possessions. The more of a good A possesses, the lower is B's utility:

$$UB = XB^{\wedge}(2/3)^{*}YB^{\wedge}(1/3) - (XA^{\wedge}(1/4) * YA^{\wedge}(1/4)). \qquad (15.12)$$

Does envy affect the equilibrium distribution of the goods?

How does the exchange process change if a large number of people want to trade part of their endowments for the same good offered by one person? This is the basic setting for the auction game discussed in the following section. We will assume that the participants in trade attempt to maximize the utilities they can derive from their endowments of goods. In Part VI of this book, we will return to the utility-maximizing consumers and introduce explicitly the effects the consumption history has on utility.

Note that the equilibrium levels depend on the initial conditions. Show this by transferring five time units from A to B and rerun the model. What does this outcome say about the prospects for fair trade?

15.4   Barter Economy Model Equations

```
XA(t) = XA(t - dt) + (- ΔX) * dt
INIT XA = 30  {Units of X}
OUTFLOWS:
ΔX = IF ΔUA >= 0 THEN EX ELSE -EX  {Units per Time Period}
XB(t) = XB(t - dt) + (ΔX) * dt
INIT XB = 5  {Units of X}
INFLOWS:
ΔX = IF ΔUA >= 0 THEN EX ELSE -EX  {Units per Time Period}
YA(t) = YA(t - dt) + (ΔY) * dt
INIT YA = 15  {Units of Y}
INFLOWS:
ΔY = IF ΔUB >= 0 THEN EY ELSE  -EY {Units per Time Period}
YB(t) = YB(t - dt) + (- ΔY) * dt
INIT YB = 40  {Units of Y}
OUTFLOWS:
ΔY = IF ΔUB >= 0 THEN EY ELSE  -EY {Units per Time Period}
CHANGE_UA = ΔUA/UA
CHANGE_UB = ΔUB/UB
EX = (.25+.1*CHANGE_UA)
EY = (.25+.1*CHANGE_UB)
UA = (XA^(1/3) * YA^(2/3))/(XA+YA)
UB = (XB^(2/3) * YB^(1/3))/(XB + YB)
ΔUA = UA - DELAY(UA,DT,25)
ΔUB = UB - DELAY(UB,DT,25)
```

15.5   Sealed-Bid, Second-Price Auction Game

In this section we describe and model a sealed-bid, second-price auction game, which is also called the *Vickrey Auction* (for details see, e.g., Shogren 1993; Shogren et al. 1994, and for the original model see Vickrey 1961). Assume that four restaurants buy food items from a supplier in a market in which they bid for the food. The restaurants are the bidders in our market and

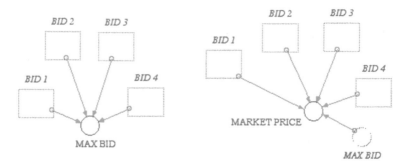

Fig. 15.6 Bid generation

are told that the food they intend to buy may contain salmonella at the same probability as found in food from other "normal" suppliers. Each of the bidders attempting to buy food for its restaurant has the choice to bid for the food on the market or buy a special sanitary food as an alternative. The bid is the differential value between the two types of food, sometimes called the *ex ante willingness to pay*.

The auction is a silent one in which only the second highest bid is known to all bidders. The second highest bid is the MARKET PRICE, the signal to which bidders may respond. For the four bids BID 1, ..., BID 4 by our restaurants and a highest bid MAX BID, the market price is calculated as

MARKET PRICE = IF MAX BID = BID 1 THEN MAX(BID 2, BID 3, BID 4)
ELSE IF MAX BID = BID 2 THEN MAX(BID 1, BID 3, BID 4)
ELSE IF MAX BID = BID 3 THEN MAX(BID 1, BID 2, BID 4)
                            ELSE MAX(BID 1, BID 2, BID 3).

$$(15.13)$$

The bids (Fig. 15.6) are generated in our model by three different types of bidders: two bidders (1 and 2) who try in slightly different ways to catch up with the market signal and overbid just enough to be the high bidder; a third bidder who ignores market signals; and one bidder (bidder number 4) who is largely random. Each bidder starts with a bid that is based on its own private risk perception. We call this the *self-reliance probability* (SRP). Bidders are assumed to assign their own weights to SRP and the market probability Q. The self-reliance probabilities are SRP1 = 0.9, SRP2 = 0.4, SRP3 = 0.1, and SRP4 = 0.5. Weights are set for the four bidders as ALPHA 1 = ALPHA 4 = 0.5, ALPHA 2 = 0.1, and ALPHA 3 = 0.9. The market probability is the same for each bidder and specified as a graph. The introduction of SRP and Q enable us to easily devise an array of bidder behaviors to describe those behaviors identified in experiments.

We defined the market history as BID$(t-1)$−MARKET PRICE$(t-1)$. We could include all the previous bids (bidder or market in a discounted difference). The problem is to transfer this value into the relative market risk perception, Q.

Fig. 15.7 Own vs. market
risk perception

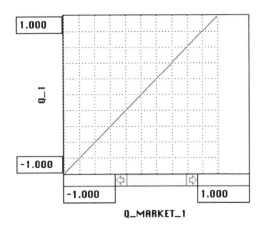

We do this with a graph of Q for each bidder. It is not likely to be a straight-line relationship, and yet we can use this line as the way to begin a calibration procedure for the model.

The equations for the initial and subsequent bids are Bayesian:

$$\text{BID } 1 = V1^*(1 - \text{ALPHA } 1^*\text{SRP } 1/(\text{ALPHA } 1 + \text{BETA } 1)) \qquad (15.14)$$

and

$$\text{BID } 2 = V2^*(1 - \text{ALPHA } 2^*\text{SRP } 2/(\text{ALPHA } 2 + \text{BETA } 2)), \qquad (15.15)$$

with $V1 = 1$ and $V2 = 2$ as the monetary values placed by the two bidders on good health. ALPHA and BETA are the coefficients that determine the reliance the bidder places on its own risk perception vs. the market risk perception, respectively. Alpha + Beta = 1, in our model (Fig. 15.7).

The initial bids for the other two bidders are specified analogously as

$$\text{BID } 3 = V3^*(1 - \text{ALPHA } 3^*\text{SRP } 3/(\text{ALPHA } 3 + \text{BETA } 3)) \qquad (15.16)$$

and

$$\text{BID } 4 = V4^*(1 - \text{ALPHA } 4^*\text{SRP } 4/(\text{ALPHA } 4 + \text{BETA } 4)). \qquad (15.17)$$

Updates to the bids are made based on the prior bids held by the bidders and reflect their individual bidding behavior:

$$\text{PRIOR } 1 = \text{IF BID } 1 \leq 1.25^*V1 \text{ THEN IF BID } 1 = \text{MAX BID}$$
$$\text{THEN } - V1^*\text{BETA } 1^*Q1/(\text{ALPHA } 1 + \text{BETA } 1)$$
$$\text{ELSE } - V1 * \text{BETA } 1^*Q1/(\text{ALPHA } 1 + \text{BETA } 1) + \text{RANDOM}(0, .1)$$
$$\text{ELSE BID } 1 \leq 1.25^*V1,$$

$$(15.18)$$

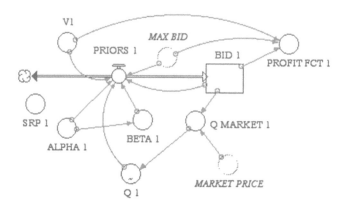

Fig. 15.8 Calculation of bid 1

PRIOR 2 = IF BID 2 ≤ 1.25*V2 THEN IF BID 2 = MAX BID
                THEN − V2*BETA 2*Q2/(ALPHA 2 + BETA 2)
                ELSE − V2*BETA 2*Q2/(ALPHA 2 + BETA 2) + RANDOM(0, .05)
                ELSE BID 2 ≤ 1.25*V2,

(15.19)

PRIOR 3 = IF BID 3 ≤ 1.25*V3 THEN IF BID 3 = MAX BID
                THEN − V3*BETA 3*Q3/(ALPHA 3 + BETA 3)
                ELSE − V3*BETA 3*Q3/(ALPHA 3 + BETA 3) + RANDOM(0, .05)
                ELSE BID 3 ≤ 1.25*V3,

(15.20)

PRIOR 4 = IF BID 4 = MAX BID
                THEN − V4*BETA 4*Q4/(ALPHA 4 + BETA 4) + RANDOM(−.5, .3)
                ELSE − V4*BETA 4*Q4/(ALPHA 4 + BETA 4) + RANDOM(−.5, .3)

(15.21)

The PRIORS are calculated using these equations. The PRIORS contain a random element, biased upward, that causes the bidder to bid slightly higher if it did not win the last bidding round. Remember, only the winner knows both the first and second highest bids. The module set up to calculate BID 1 is shown in Fig. 15.8 for the case of the first bidder. The modules for the other three bidders are specified analogously.

The profit functions for each of the bidders are relative net value of the food bought if the bidder did make the maximum bid and 0 otherwise. For example, for the first bidder the profit function is

PROFIT FCT 1 = IF BID 1
                      = MAX BID THEN(V1 − BID 1)/V1 ELSE 0.   (15.22)

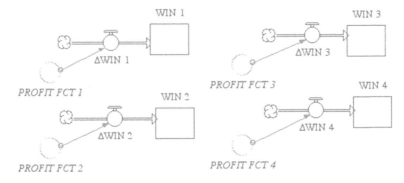

Fig. 15.9 Accumulation of winning bids

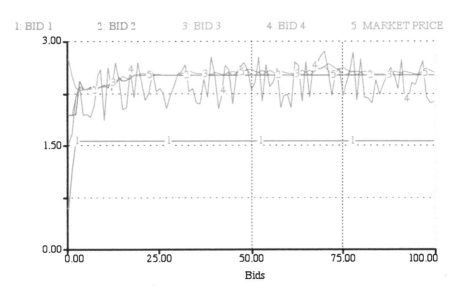

Fig. 15.10 Bid dynamics

To see who wins the bidding contest for the food for its restaurant, we accumulate the winning bids of each round of bids in stocks WIN 1, . . . , WIN 4 (Fig. 15.9).

The model shows that the first bidder is bid out of the market first and that the third bidder soon wins the auction. The resulting market price ultimately fluctuates around $2.60 per food item, but it keeps changing because the fourth bidder, whose behavior is largely random, makes an occasional bid (Figs. 15.10–15.12).

Can you find other behavioral assumptions and integrate them in the auction game? What are the results for an increasing number of bidders that attempt to outbid each other?

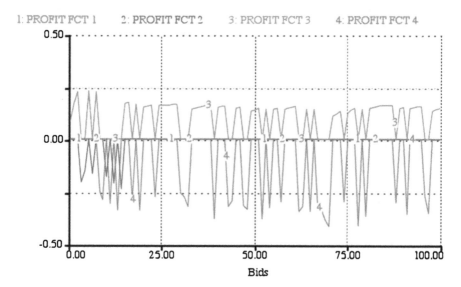

Fig. 15.11 Profits

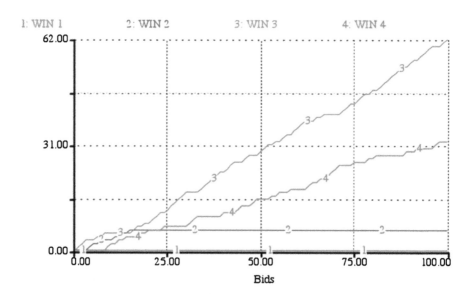

Fig. 15.12 Wins

Economic game theory is a very powerful analytic tool in describing economic outcomes, especially when coupled with a simulation procedure. Here we have managed to model the behavior of four different bidders—behavior actually found in related experiments. It seems to us that, when economic simulation games are based on experimental data, economics becomes a true science.

15.6   Sealed-Bid Second-Price Auction Game

BID_1(t) = BID_1(t - dt) + (PRIORS_1) * dt
INIT BID_1 = V1*(1-ALPHA_1*SRP_1/(ALPHA_1+BETA_1))
{Dollars per Food Item}
INFLOWS:
PRIORS_1 = IF BID_1≤1.25*V1 THEN IF BID_1=MAX_BID
THEN -V1*BETA_1*Q_1/(ALPHA_1+BETA_1) ELSE -

V1*BETA_1*Q_1/(ALPHA_1+BETA_1)+RANDOM(0,.1)
ELSE BID_1≤1.25*V1   {Dollars per Food Item per Bid}

BID_2(t) = BID_2(t - dt) + (PRIORS_2) * dt
INIT BID_2 = V2*(1-ALPHA_2*SRP_2/(ALPHA_2+BETA_2))
{Dollars per Food Item}
INFLOWS:
PRIORS_2 = IF BID_2≤1.25*V2 THEN IF BID_2=MAX_BID
THEN -V2*BETA_2*Q_2/(ALPHA_2+BETA_2) ELSE -
V2*BETA_2*Q_2/(ALPHA_2+BETA_2)+RANDOM(0,.05)
ELSE BID_2≤1.25*V2 {Dollars per Food Item per Bid}

BID_3(t) = BID_3(t - dt) + (PRIORS_3) * dt
INIT BID_3 = V3*(1-ALPHA_3*SRP_3/(ALPHA_3+BETA_3))
{Dollars per Food Item}
INFLOWS:
PRIORS_3 = IF BID_3≤1.25*V3 THEN IF BID_3=MAX_BID
THEN -V3*BETA_3*Q_3/(ALPHA_3+BETA_3) ELSE -
V3*BETA_3*Q_3/(ALPHA_3+BETA_3)+RANDOM(0,.05)
ELSE BID_3≤1.25*V3 {Dollars per Food Item per Bid}

BID_4(t) = BID_4(t - dt) + (PRIORS_4) * dt
INIT BID_4 = V4*(1-ALPHA_4*SRP_4/(ALPHA_4+BETA_4))
{Dollars per Food Item}
INFLOWS:
PRIORS_4 = IF BID_4=MAX_BID
THEN -V4*BETA_4*Q_4/(ALPHA_4+BETA_4)+RANDOM(-
.5,.3) ELSE
-V4*BETA_4*Q_4/(ALPHA_4+BETA_4)+RANDOM(-.5,.3)
{Dollars per Food Item per Bid}

WIN_1(t) = WIN_1(t - dt) + (ΔWIN_1) * dt
INIT WIN_1 = 0
INFLOWS:
ΔWIN_1 = if PROFIT_FCT_1 ≠ 0 then 1 else 0

WIN_2(t) = WIN_2(t - dt) + (ΔWIN_2) * dt
INIT WIN_2 = 0
INFLOWS:
ΔWIN_2 = if PROFIT_FCT_2 ≠ 0 then 1 else 0

WIN_3(t) = WIN_3(t - dt) + (ΔWIN_3) * dt
INIT WIN_3 = 0
INFLOWS:
ΔWIN_3 = if PROFIT_FCT_3 ≠ 0 then 1 else 0

WIN_4(t) = WIN_4(t - dt) + (ΔWIN_4) * dt
INIT WIN_4 = 0
INFLOWS:
ΔWIN_4 = if PROFIT_FCT_4 ≠ 0 then 1 else 0

ALPHA_1 = .5
ALPHA_2 = .1
ALPHA_3 = .9
ALPHA_4 = .5
BETA_1 = 1-ALPHA_1
BETA_2 = 1-ALPHA_2
BETA_3 = 1-ALPHA_3
BETA_4 = 1-ALPHA_4
MARKET_PRICE = IF MAX_BID=BID_1 THEN
MAX(BID_2,BID_3,BID_4) ELSE IF MAX_BID=BID_2 THEN
MAX(BID_1,BID_3,BID_4) ELSE IF MAX_BID = BID_3 THEN
MAX(BID_1,BID_2,BID_4) ELSE MAX(BID_1,BID_2,BID_3)
MAX_BID = MAX(BID_1,BID_2,BID_3,BID_4) {Dollars per
food item}
PROFIT_FCT_1 = IF BID_1=MAX_BID THEN (V1-BID_1)/V1
ELSE 0 {Dollars per Dollar}
PROFIT_FCT_2 = IF BID_2=MAX_BID THEN (V2-BID_2)/V2
ELSE 0 {Dollars per Dollar}
PROFIT_FCT_3 = IF BID_3 = MAX_BID THEN (V3-
BID_3)/V3 ELSE 0 {Dollars per Dollar}
PROFIT_FCT_4 = IF BID_4 = MAX_BID THEN (V4-
BID_4)/V4 ELSE 0 {Dollars per Dollar}
Q_MARKET_1 = BID_1-MARKET_PRICE
Q_MARKET_2 = BID_2-MARKET_PRICE
Q_MARKET_3 = BID_3-MARKET_PRICE

Q_MARKET_4 = BID_4-MARKET_PRICE
SRP_1 = .9
SRP_2 = .4
SRP_3 = .1
SRP_4 = .5
V1 = 1 {Dollars per Food Item}
V2 = 2 {Dollars per Food Item}
V3 = 3 {Dollars per Food Item}
V4 = 2 {Dollars per Food Item}
Q_1 = GRAPH(Q_MARKET_1)
(-1.00, -1.00), (-0.8, -0.8), (-0.6, -0.6), (-0.4, -0.4), (-0.2, -0.2),
(-6.66e-17, 0.00), (0.2, 0.2), (0.4, 0.4), (0.6, 0.6), (0.8, 0.8), (1,
1.00)
Q_2 = GRAPH(Q_MARKET_2)
(-1.00, -1.00), (-0.8, -0.8), (-0.6, -0.6), (-0.4, -0.4), (-0.2, -0.2),
(-6.66e-17, 0.00), (0.2, 0.2), (0.4, 0.4), (0.6, 0.6), (0.8, 0.8), (1,
1.00)
Q_3 = GRAPH(Q_MARKET_3)
(-1.00, -1.00), (-0.8, -0.8), (-0.6, -0.6), (-0.4, -0.4), (-0.2, -0.2),
(-6.66e-17, 0.00), (0.2, 0.2), (0.4, 0.4), (0.6, 0.6), (0.8, 0.8), (1,
1.00)
Q_4 = GRAPH(Q_MARKET_4)
(-1.00, -1.00), (-0.8, -0.8), (-0.6, -0.6), (-0.4, -0.4), (-0.2, -0.2),
(-6.66e-17, 0.00), (0.2, 0.2), (0.4, 0.4), (0.6, 0.6), (0.8, 0.8), (1,
1.00)

References

Hanneman RA (1988) Computer-assisted theory building: modeling dynamic social systems. Sage, Newbury Park

Shogren JF (1993) Experimental markets and environmental policy. Agr Resource Econ Rev 12:117–129

Shogren JF, Shin SY, Hayes DJ, Kliebenstein JB (1994) Resolving differences in willingness to pay and willingness to accept. Am Econ Rev 84:255–270

Vickrey W (1961) Counterspeculation, auctions, and competitive sealed tenders. J Finance 16:8–37

Part IV
Modeling Optimal Use
of Nonrenewable Resources

Chapter 16
Competitive Scarcity

For the Poorer fort, ... Horfe-dung in Balls with Saw-duft,
or the duft of Smalcoale, or Charcoale duft, dryed,
is good Fewell, but the fmell is offenfive

(Richard Gesling, Ingineer, *Artificiall Fire, or*
Coale for Rich and Poor, A Recipe for Making
Briquettes, 1644. British Museum, 669, f. 10(11).

16.1 Basic Model

In the previous chapters on the models of firms, we used STELLA to find the profit-maximizing input and output quantities at a point in time. We ran the models over time as if the firm were able to "feel its way towards the optimum." Similarly, the auction game was assumed to take place over time, but the decision making itself did not take into account the time over which the decision is made. In contrast to these models, we now turn our attention to the problem of finding an *optimal path* through time. Specifically, we are interested in how much of a resource to extract at each point in time over the lifetime of a finite, nonrenewable resource.

As we show in this and the following chapters, the seemingly complex problem of determining a whole series of optimal points along a path—rather than just a single optimal point, as we have done, for example, in Chaps. 8, 11, and 12—can be simplified significantly. We will first show some of the general features of this path. This part of the problem is solved with analytical methods. Once features of the optimal path are identified, the remaining problem is to ensure that the system of interest—in our case, the resource-extracting firm—positions itself at the outset in

A save-disabled version of STELLA® and the computer models of this book are available at www.iseesystems.com/modelingeconomicsystems.

M. Ruth and B. Hannon, *Modeling Dynamic Economic Systems,*
Modeling Dynamic Systems, DOI 10.1007/978-1-4614-2209-9_16,
© Springer Science+Business Media, LLC 2012

one point on the path. The second part of this problem is solved through trial and error with a STELLA model that incorporates the analytical solution for the optimal trajectory of the system.

Once the initial point on the optimal path of the system is identified, the whole system behaves optimally over time. The dynamic optimization problem, therefore, boils down to the optimal choice for a single period, the starting point for our model. Recognize that this is a mechanistic approach to understanding the dynamics of a system. In Part VI of this book, we contrast this mechanistic view with one in which prediction of a system's future behavior is impossible based on knowledge of its past and current states. For examples of purely simulation-based approaches to the issue of optimal extraction of a nonrenewable resource, see Shelden and Hassler (1980) or Shelden (1982).

Assume that we have an oil field. Oil is not a renewable resource. When the reservoir under our field is depleted, the wells will run dry. Assume further that the size of the reservoir is perfectly known and its quality is uniform. Therefore, we want to maximize the current value of profits from selling our crude oil. As previously, we assume perfect competition in the marketplace; the price we can ask for our crude equals the OPEC price in each period. Crude oil is unique; there are no effective substitutes for it, and it is consumed when it is used—no recycled oil. To keep the model simple, we assume that no substitute will be discovered and that our costs of extraction will remain the same. We also ignore all government subsidies, depletion allowances, and price supports.

The reservoir under our oil field holds a finite amount of petroleum. We control the rate at which our oil is sold to a refinery. If we deplete our reservoir, we can invest the profits we earn, for example, in another business or another potential oil field. Of course, the new business might fail or the new field might not live up to its potential. We have an alternative: we need not extract any of the oil. By thus keeping our oil off the market, we help drive up the price for crude oil. Then, if the price were high enough, we could begin pumping again. Of course, during the waiting period, some of our costs will continue and we would lose the present use of the profits. Which is the better choice? The optimal behavior seems to lie between the two extremes: full extraction now or no pumping until the price goes up.

Optimal behavior, in the sense of maximizing the present value of cumulative profit, would be for all oil producers to sell their crude at a rate that would continuously raise its price. Such thinking led to the Organization of Petroleum Exporting Countries (OPEC). As oil producers, we would begin by selling a large amount of crude. This would give us a high present value of profits. The amount of oil in the reservoir is reduced by the amount we have extracted, and this would be the case for others who followed the same plan. Therefore, less crude oil would be available in the future. This lower supply would lead to higher prices along the demand curve. With a higher price for our crude, and less of it in the reservoir, we would pump less oil in future periods.

The problem becomes one of choosing an optimal schedule of production. The best way to solve such a complex problem is to use analytic techniques to derive the governing dynamic equations (differential equations) and then solve them numerically with the aid of STELLA. Most often, such equations are not analytically solvable. So we combine analytic and numerical techniques to find the optimal path. We will discover though that this is not all we need. The optimal solution depends on several initial and boundary conditions, and these are met by repeated trial runs in STELLA. There are some strategies that we can use to reduce the drudgery. We begin with the analytic development.

The problem of finding the optimal path for the extraction of a nonrenewable resource can be formalized as follows. The objective of our petroleum extraction is to maximize the cumulative stream of current value profits discounted each period at an interest rate I, that is,

$$Maximize\ CPVP = \int_0^T CVPe^{-I^*t}dt = \int_0^T (P^*Q - C)e^{-I^*t}dt. \qquad (16.1)$$

We want to maximize these profits over a period that begins today ($t = 0$) and continues until period T, the time when the reservoir is totally depleted. Our model will determine T, of course. Q represents output, the variable we can control to maximize our profits. In (16.1), P is the price of crude oil in period t and C is the cost of extracting the crude oil. The cost of extraction is a function of the output:

$$C = B^*Q^{\wedge}DELTA, \qquad (16.2)$$

where B and DELTA are constants. Basically, our production costs rise as we pump more oil. In (16.1), I is the rate of discount, whose value is known and remains constant from one period to the next.

The amount of oil in the reservoir is finite: no "stream" feeds it and we cannot wait for more oil to be "produced" naturally. This places a constraint on our maximization: we cannot extract more oil than is there now. Our initial stock of oil must be greater than or equal to the amount we can pump in T periods:

$$Y(t = 0) = \int_0^T Qdt. \qquad (16.3)$$

We recognize that the oil is a nonrenewable resource (i.e., the oil will not be replaced in the reservoir) by stating

$$Q \geq 0. \qquad (16.4)$$

Finally, we want to show that the change in the amount of oil in the reservoir, $\dot{Y} = \partial Y/\partial t$, reflects only the amount of oil pumped; no other factor changes the reservoir amount. We state this formally as

$$\dot{Y} = -Q. \qquad (16.5)$$

$\dot{Y}$ is negative because the amount decreases as we pump out oil.

We are now ready to determine the optimal rate of extraction Q in time period t. We do this by maximizing the function H, called the Hamiltonian, relative to Q, which is our control variable:

$$H = (P^*Q - C)^* e^{-I^*t} \lambda^* Q, \tag{16.6}$$

where $(P * Q - C)$ is the current value of profits at some period. It is discounted at the interest rate I to yield the present value of profits in that period. The term $\lambda * Q$ reflects the "penalty" (λ) for reducing our supply of oil in the reservoir by an amount Q. Another way to interpret the term is this: λ represents what a petroleum producer would pay to have one more unit of oil in the reservoir; we call this the (discounted) *scarcity rent rate*.

The method of maximizing our profits over time is known as the *optimal control theory* (Dorfman 1969). We will assume that our oil operation is competitive and $P = P(t)$ is not a function of Q. The optimal extraction conditions (Kamien and Schwartz 1983) for our oil operation are given by the partial derivative of the Hamiltonian relative to Q, our control variable,

$$\frac{\partial H}{\partial Q} = \left[P - \frac{\partial C}{\partial Q} \right]^* e^{-I^*t} - \lambda = 0, \tag{16.7}$$

and the way H is affected by a change in the amount of oil left in the reservoir, Y,

$$-\frac{\partial H}{\partial Y} = \dot{\lambda}, \tag{16.8}$$

where the dot on λ refers to the partial derivative with respect to time. Since H is not a function of Y, we know

$$\dot{\lambda} = 0. \tag{16.9}$$

Equation (16.7) yields

$$\lambda = (P - MC)^* e^{-I^*t}. \tag{16.10}$$

Taking the first partial derivative with respect to time in (16.10) and combining it with (16.9) helps us to eliminate λ from the system of equations:

$$\dot{\lambda} = (\dot{P} - \dot{MC})^* e^{-I^*t} - I^* \lambda. \tag{16.11}$$

Inserting (16.10) into (16.11) enables us now to express changes in price of the resource over time solely in terms of observable quantities:

$$\dot{P} = I^*(P - MC) + \dot{MC}. \tag{16.12}$$

Recognize that (16.12) can also be rewritten as

$$\frac{\dot{P} - \dot{MC}}{P - MC} = I. \tag{16.13}$$

Since $P - MC = \lambda^* e^{I * t}$ is the undiscounted scarcity rent rate, let us call it μ; we have

$$\dot{\mu}/\mu = I. \tag{16.14}$$

Equation (16.14) describes a condition called the *Hotelling rule* (Hotelling 1931). That is, for extraction of a nonrenewable resource to be optimal (to maximize the present value of profits), the undiscounted scarcity rent rate of the resource must rise at the rate of interest. This means that the *undiscounted* scarcity rent rate, μ/μ, must equal the interest rate we could garner from investing our profits.

When we discussed the maximum profit of a manufacturer in a perfectly competitive market, we found that the price equals the marginal cost of producing its good. Here, because we are dealing with a nonrenewable resource, we must add to that marginal cost a rent, λ, to represent the irreversible loss of the resource, which explains the term *scarcity rent rate*. The relationship between the solutions of the profit maximization for the extraction of a nonrenewable resource and the standard competitive equilibrium are shown in Fig. 16.1. The "endowment rent" shown in the figure is realized by well-endowed (low-cost) producers, such as the ones we discuss in Chap. 14.

In (16.12), we showed how we could maximize our profits by following a time path in pumping observable amounts of oil from our reservoir. Now, we must express these conditions in a STELLA model to learn exactly how much oil we can pump in each period.

Recognize that (16.12) is used in the STELLA model to calculate the change in price over time. It is specified as a flow ΔP into the stock P. The integration is done in STELLA by multiplying this flow by DT:

$$P(t) = P(t - dt) + (\Delta P)*DT \tag{16.15}$$

with

$$\Delta P = I^*(P - MC) + \Delta MC. \tag{16.16}$$

The value ΔMC takes on in our model is dependent on the length of the time step used for the integration. Hence, we need to correct ΔMC for the length of the time step DT. This correction can easily be made by dividing ΔMC by DT:

$$\Delta P = I^*(P - MC) + \Delta MC/DT. \tag{16.17}$$

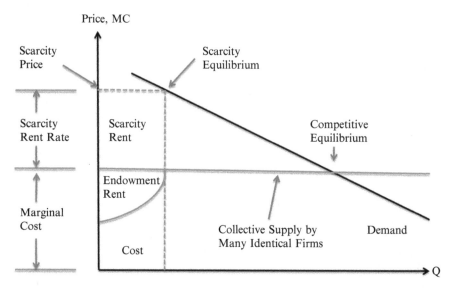

Fig. 16.1 Price, marginal cost, and scarcity rent rate in perfectly competitive nonrenewable resource markets

This is the expression for ΔP that will be used in the model.

The STELLA model (Fig. 16.2) also requires that we specify marginal cost and the change in marginal cost over time. For the cost function of (16.2), the marginal cost function is

$$MC = DELTA^*B^*Q^\wedge(DELTA - 1) \qquad (16.18)$$

and the change in marginal cost over time is calculated in the STELLA program as

$$\Delta MC = MC - DELAY(MC, DT). \qquad (16.19)$$

Finally, let us assume for simplicity that there are N identical resource extracting firms on the market that compete with each other. Each of those firms supplies one Nth of the demand for the nonrenewable resource. Total demand is given by the demand function

$$P = 10 - 0.00338\, Q, \qquad (16.20)$$

which we can solve for the demand that is met by each of the N firms:

$$Q = \frac{(10 - P)/0.00338}{N}. \qquad (16.21)$$

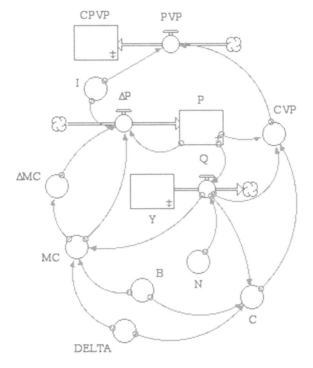

Fig. 16.2 Optimal extraction from a nonrenewable resource

For simplicity we assume that $N = 10$ in this and the following problems of competitive resource extraction. Also, assume that the discount rate is fixed at 5%.

Because we are one of ten identical firms who own the reservoir, we know how much oil is in it (at least, we think we do). Let us assume that the amount in all reservoirs combined is 60,000 units of oil. Hence, we own $60,000/N = 6,000$ units of it.

Note when running this model, there are some terminal conditions you must meet. These are called the transversality conditions and are needed to insure optimality, in addition to the necessary conditions that lead to (16.17). The transversality conditions for this type of problem are as follows: $\lambda(T) \geq 0$, $\lambda(T) * Y(T) = 0$, where T would be the terminal time, when either Q or Y became 0, or, T is infinity, in the case where both Q and Y asymptotically approached 0.

When the finite resource is of uniform quality, we know that all of it will be used, resulting in a finite terminal time. From (16.7) and (16.9) we can see that l will be a positive constant for all times, so $\lambda(T) > 0$ and therefore $Y(T)$ must be 0 to satisfy the second condition. So we pick an initial price that will bring Y to 0 such that a slightly higher price will leave some resource in the ground. This $P0$ is the highest initial price that still reduces the final amount of resource to 0. It can be observed

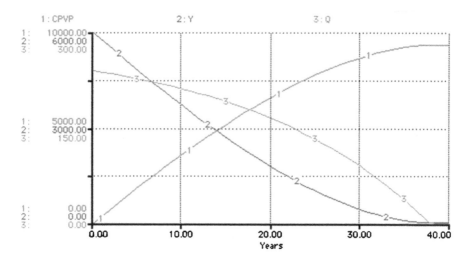

Fig. 16.3 Profit-maximizing behavior of nonrenewable resource extracting firms

that this condition produces the highest CPVP of all such initial prices. $Q(T)$ must also be 0, since no Q can exist without a resource, and if Q is 0 before Y is 0, Y will never reach 0. The exact terminal time T is discovered by this process. This line of reasoning holds for all our optimization models where the cost of extraction is only a function of Q.

Note that the cost function, the number of firms, the demand curve, the initial amount of remaining resource and the firms discount rate are required data for this model. But our differential (16.17) and the terminal conditions guide us along the optimal path to depletion. The key to this solution is the assumption of the firm maximizing behavior in (16.1).

Choose a different number N of firms in the market. Set DT $= 1$ and run the model several times to find the optimal initial value of P. How does N affect CPVP?

Figure 16.3 shows the time path that maximizes the cumulative present value of profits for the number of identical firms $N = 10$. Note the smooth decline in the production rate, Q; a steady rise in price, P; and a fall in output Y. We see that the production rate is 0 just as the choke-off price is reached. A bit of the oil is left in the reservoir. Decrease the discount rate, reoptimize the model, and see the effect on the amount of reserves left in the ground.

Set up the model to start with the terminal conditions and run the model backward. This should give you the relationship between the different reserve sizes and the corresponding optimal price. Run the model until $Y = 6,000$ and determine the starting price $P(t = 0)$. Suppose we stop the problem at $t = 20$ and reoptimize on the remaining reserve? What would happen?

You can use these trial and error techniques to assure yourself that the discounted scarcity rent rate really does rise at the rate of interest, which is what the analytic theory indicates it should in the absence of any stock effects. Furthermore, you will notice that CPVP is maximized when the choke-off price is reached by a production rate that smoothly goes to 0 at the same time.

Realize that these firms were probably coasting along at equilibrium, pumping their resources away, at the time they first realized they were dealing with a finite resource. If they were at equilibrium, CVP was 0 (it is not in this chapter), and AC = MC = P. The simple cost function used here will not allow this condition to be established (see Chap. 14, Sect. 14.1). So, change the cost function to $C = B*Q^{0.5} + DELTA*Q^2$, with $B = 5$, and $DELTA = 0.02$. Form the variable AC and experiment with the number of firms until MC = AC at the start of the model. This is the prescarcity equilibrium price. The model will show this initial postscarcity price, along with the pre- and postscarcity MC and AC (equal at $t = 0$). Then determine the initial price that meets the boundary conditions, then go back and adjust the number of firms, etc. until the boundary conditions are met and the AC = MC at $t = 0$. What is the prescarcity initial price, the number of firms and initial scarcity price? Note that from $t = 0$ on in this model, MC and P differ by the scarcity rent. So here is what the firms are doing. The instant before they implement the scarcity behavior, $P = MC = AC$ and all these firms are at equilibrium. The instant that the scarcity behavior begins, the price jumps to contain the scarcity rent, AC begins to slowly rise, MC begins to slowly decline.

The model for maximizing CPVP is rather intuitive and instructive, but it cannot be applied to all cases. In the model we assumed that the cost of extracting oil from the reservoir did not change until very little was left. But, what if the reservoir were uneven? Only some of the oil would be found and extracted easily and the cost of extracting the rest would have to include increasing costs of extracting the resource, say, from greater depths. We have seen, albeit in a very simplified way, the consequences of a depletion effect in the case of uneven resource endowments in Chap. 3, where declines in ore grade lead to an increase in the energy necessary to extract a mineral from the ground. The model developed here would not account for these differences and the interest rate would rise faster than the scarcity rent rate. This means that less of the resource might be used. We augment our little model from the theory and enrich it in the following sections to cover a number of different cases. Among those extensions are nonlinear demand curves, sudden demand shifts, gradually increasing availability of a perfect substitute, depletion effects on the cost of extraction, and more. First, however, we need to understand the effects the numerical solution technique has on the results of our model. This is done in the following section.

16.2   Basic Competitive Scarcity Model Equations

CPVP(t) = CPVP(t - dt) + (PVP) * dt
INIT CPVP = 0 {The cummulative present value of the profit; Dollars}
INFLOWS:
PVP = CVP/(1+I)^TIME {Dollars per Year}
P(t) = P(t - dt) + (ΔP) * dt
INIT P = 1.89
INFLOWS:
ΔP = I*(P - MC) + ΔMC/dt {Tons per Time Period}
Y(t) = Y(t - dt) + (- Q) * dt
INIT Y = 60000/N {Units of X}
OUTFLOWS:
Q = (10 - P)/.00338/N   {Tons per Time Period}
B = .1
C = B*Q^DELTA {This is the total cost and it is independent of any stock size or interest rate. Dollars per Year}

CVP = P*Q-C {Dollars per Year}
DELTA = 1.2   {This value must be greater than one so that the C vs Q curve is convex.}
I = .05 {Dollars per Dollar Invested per Time Period}
MC = B*DELTA*Q^(DELTA-1)
N = 10
ΔMC = (MC- DELAY(MC,DT)) {Dollars per Unit of X}

16.3   Competitive Scarcity with Various DT

Let us take the model of the competitive firm dealing with a finite resource and reduce the DT below 1 or choose the Runge–Kutta-4 numerical solution technique in the TIME SPECS menu. Not only is the calculation of the optimal time path slowed by the increased precision of the numerical solution method, but we also get a different answer for the initial price that produces the maximum cumulative present value of profits. Using the analytic form for MC directly for the specification of these converters reduces inaccuracies that stemmed from the numerical solution. In this, as in the previous problem, that substitution has been made (Fig. 16.4).

Even though we make use here of the analytical solution of the problem, the answer is still sensitive to the choice of DT and the calculation form. Run this problem with DT = 0.125 and 1.0 and with the Euler and Runge–Kutta-4 solution forms to see the variation in the answer.

Figure 16.5 shows the optimal time path for DT = 0.125. The profit-maximizing initial price is $P = 1.83$ with a maximum CPVP = 9,098, which deviates slightly from the result in the previous section. There, we had an optimal initial price $P = 1.89$ and a maximum CPVP = 9,376. When do we know that we have a sufficiently accurate solution to our model? Is DT = 0.125 with Runge–Kutta-4 far enough to go in the search for accuracy?

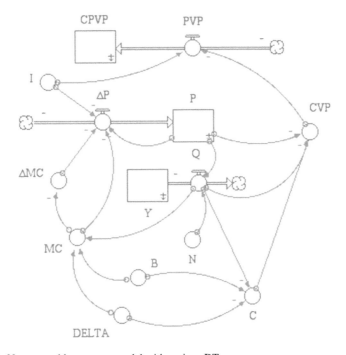

Fig. 16.4 Nonrenewable resource model with various DT

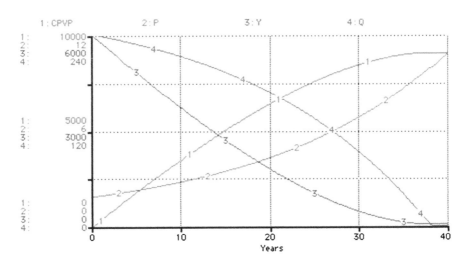

Fig. 16.5 Optimal extraction in the nonrenewable resource model with various DT

Modify this model to deal with the following situation. The demand curve is linear for the first 20 periods. Use the specification of the demand curve shown in the previous section. After period 20, a new, nonlinear demand curve is the relevant one. How do the optimal initial price and the resulting time paths change? Can you change this model to have a series of linear demand curves for, say, ten periods each, whose slope is gradually decreasing as the choke-off price is approached?

Run the model with a demand curve that does not have a choke-off price. Try $P = 6,000/Q-2$. Now all the resource will be exhausted.

16.4 Competitive Scarcity with Various DT Model Equations

CPVP(t) = CPVP(t - dt) + (PVP) * dt
INIT CPVP = 0 {The cummulative present value of the profit; Dollars}
INFLOWS:
PVP = CVP/(1+I)^TIME {Dollars per Year}
P(t) = P(t - dt) + (ΔP) * dt
INIT P = 1.83 {Dollars per Units of X}
INFLOWS:
ΔP = I*(P - MC) + ΔMC {Tons per Time Period}
Y(t) = Y(t - dt) + (- Q) * dt
INIT Y = 60000/N {Units of X}
OUTFLOWS:
Q = (10 - P)/.00338/N {Tons per Time Period}
B = .1
C = B*Q^DELTA {This is the total cost and it is independent of any stock size or interest rate. Dollars per Year}
CVP = P*Q-C {Dollars per Year}
DELTA = 1.2 {This value must be greater than one so that the C vs Q curve is convex.}
I = .05 {Dollars per Dollar Invested per Time Period}
MC = B*DELTA*Q^(DELTA-1)
N = 10
DOCUMENT: The number of identical firms.
ΔMC = (MC- DELAY(MC,DT))/DT {Dollars per Unit of X}

References

Dorfman R (1969) An economic interpretation of optimal control theory. Am Econ Rev 59:817–831
Hotelling HC (1931) The economics of exhaustible resources. J Polit Econ 39:137–175
Kamien MI, Schwartz NL (1983) Dynamic optimization: the calculus of variations and optimal control in economics and management. North-Holland, New York
Shelden RA, Hassler JC (1980) Optimal management of non-renewable resources using computer simulation. J Environ Manage 10:115–123
Shelden RA (1982) Computer simulation and the optimal management of non-renewable resources, taxation, profit maximization, and negotiation. J Environ Manage 15:131–138

Chapter 17
Competitive Scarcity with Substitution

> *We must gather and group appearances, until the scientific*
> *imagination discerns their hidden laws, and unity arises from*
> *variety; and then from unity we must rededuce variety, and*
> *force the discovered law to utter its revelations of the future.*

W. R. Hamilton

17.1   Price Effects

In the previous chapter, we modeled a simple case of nonrenewable resource extraction in the absence of a substitute. Here, we use the same kind of analysis that draws on the model's optimality and transversality conditions. But now suppose that we have a producer of a finite resource who realizes that its original version of the demand is increasingly in error—people are finding substitutes for the mineral. Perhaps it sells copper and people are shifting to aluminum as the price of copper is raised. We represent this process as a new nonlinear demand curve that is tangent to the original linear curve at the initial or starting price. From there, the new demand curve departs increasingly from the linear one. You can see this in the sketch of Fig. 17.1: the linear demand curve runs from the lower right to the upper left corners of the graph and it touches the newly revealed demand curve at its starting point. The equation for this curve is

$$Q = Q_0 + A * P^2, \tag{17.1}$$

A save-disabled version of STELLA® and the computer models of this book are available at www.iseesystems.com/modelingeconomicsystems.

M. Ruth and B. Hannon, *Modeling Dynamic Economic Systems*,
Modeling Dynamic Systems, DOI 10.1007/978-1-4614-2209-9_17,
© Springer Science+Business Media, LLC 2012

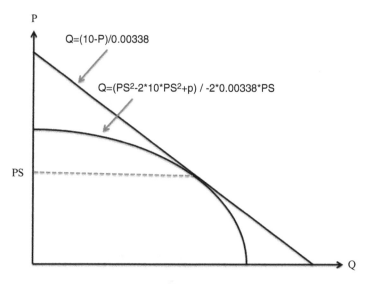

Fig. 17.1 Price effects from the emergence of substitutes for a nonrenewable resource

which, when the coefficients are solved to produce the starting tangent gives

$$Q = \frac{PS^2 - 2 * P_0 * PS + P^2}{-2 * M * PS},$$ (17.2)

where PS is the starting price, P_0 is the terminal price of the linear demand curve (10 in our model), P is the price, and M is the slope of the linear demand curve (-0.00338 in our model). Again, we divide by $N = 10$ to calculate the demand that is met by one of the ten identical firms in the market. Note how the starting price is set up in the model (Fig. 17.2).

Our producer does not know beforehand how the new curve is going to unwind. Rather, it discovers the new demand curve as the firm proceeds to extract the mineral from its mine. So how is this reaction different from the behavior in the previous section? What would you expect? If one sees with increasing clarity that the demand curve is turning toward the horizontal, one knows people are fleeing one's price. The company therefore produces an optimal path that reduces its stock faster than before, mainly because it appears that the firm has less control over its consumers. The choke-off price is just a little over $5 per unit of the mineral rather than the $10 in the previous model.

Can you construct a demand curve that is also tangent to the linear curve at the starting price but is asymptotic to the price axis? How would use of this curve change the current situation? The equation of such a curve is

$$Q = \frac{2\sqrt{P * PS} - PS - 10}{-0.00338}$$ (17.3)

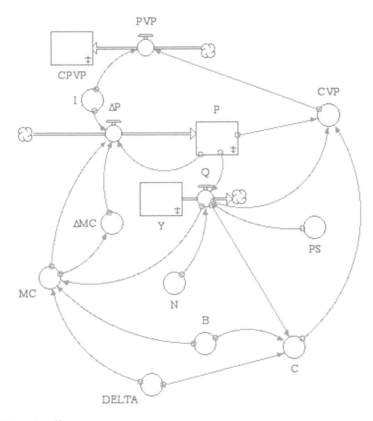

Fig. 17.2 Price effects

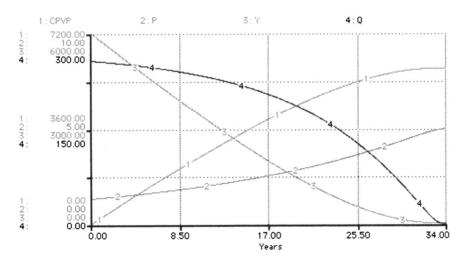

Fig. 17.3 Optimal depletion with price effect

What are the starting price and optimal time path for this demand curve? Let the number of firms decline as the demand for the resource declines. How does this affect the terminal time, the amount of resources left in the ground, and CPVP? Make an educated guess before you run the model. As in Chap. 16, the transversality condition require $Y(T) = 0$. Choose the highest price that just zeros out the final amount of resource and this will reveal the optimal path.

17.2 Competitive Scarcity with Price Effect Model Equations

```
CPVP(t) = CPVP(t - dt) + (PVP) * dt
INIT CPVP = 0{The cummulative present value of the profit }
INFLOWS:
PVP = CVP/(1+I)^TIME {Dollars per Time Period}
P(t) = P(t - dt) + (ΔP) * dt
INIT P = PS {Dollars per Unit Mineral Extracted}
INFLOWS:
ΔP = I*(P - MC) + ΔMC {Dollars per Time Period}
Y(t) = Y(t - dt) + (- Q) * dt
INIT Y = 60000/N {Units of Mineral}
OUTFLOWS:
Q = (PS^2-2*10*PS+P^2)/(-2*.00338*PS)/N {Units of Mineral Extracted Per
Time Period}
B = .1
C = B*Q^DELTA {Dollars per Time Period}
CVP = P*Q-C {Dollars per Time Period}
DELTA = 1.2 {This value must be > one so that the cost vs q curve is convex.}
I = .05 {Dollars per Dollar per Time Period}
MC = B*DELTA*Q^(DELTA-1) {Dollars per Unit Mineral Extracted}
N = 10
PS = 1.30
ΔMC = (MC - DELAY(MC,DT,MC-.000104))/DT {Dollars per Unit Mineral
Extracted per time step.}
```

17.3 Sudden Demand Shift

In the previous problem, the mine owning firm gradually discovers the "real" demand curve. The changes in demand may not be gradual but come as a shock to the system at some future time. The model of this section deals with such a sudden demand shift. In its structure, the STELLA model in Fig. 17.4 is very similar to that of the previous problem.

Let us assume that there is a price level at which substitutes suddenly are economical to produce and sell and therefore start to compete for the mineral extracted from the mine. In such a scenario, the extraction rate would be increased

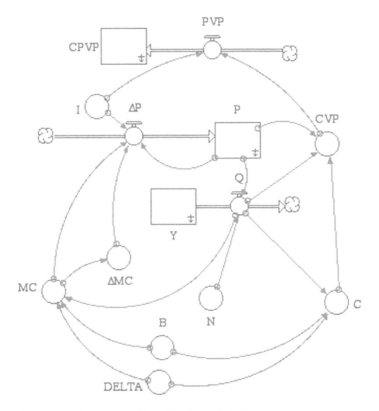

Fig. 17.4 Nonrenewable resource with sudden demand shift

once that critical price level is reached. Set this price level at $P = 5$ and have a demand shift occur from the linear demand curve

$$Q = \frac{(10 - P)/0.00338}{N} \tag{17.4}$$

to

$$Q = \frac{(7 - P)/0.00338*2}{N} \tag{17.5}$$

The transversality conditions require (see Chap. 16) that we find a price in time period zero (P0), which zeros out the resource and maximizes CPVP. That P0 is the largest one that just exhausts the resource (any larger P0 would leave some resource in situ).

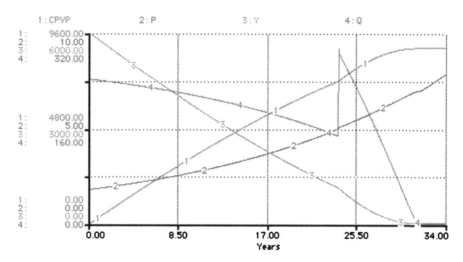

Fig. 17.5 Resource and market dynamics with sudden demand shift

The result, shown in Fig. 17.5, is a shortened time frame over which extraction takes place, a lower initial price, and a lower CPVP than if the linear demand curve of (16.1) had prevailed.

17.4   Sudden Demand Shift Model Equations

CPVP(t) = CPVP(t - dt) + (PVP) * dt
INIT CPVP = 0{The cummulative present value of the profit }
INFLOWS:
PVP = CVP/(1+I)^TIME {Dollars per Time Period}

P(t) = P(t - dt) + (ΔP) * dt
INIT P = 1.785 {Dollars per Unit Mineral Extracted}
INFLOWS:
ΔP = I*(P - MC) + ΔMC {Dollars per Time Period}
Y(t) = Y(t - dt) + (- Q) * dt
INIT Y = 60000/N {Units of Mineral}
OUTFLOWS:
Q = IF P <= 5 THEN (10 - P)/.00338/N ELSE (7 - P)/(.00338*.2)/N {Units of
Mineral Extracted Per Time Period}
B = .1
C = B*Q^DELTA {Dollars per Time Period}
CVP = P*Q-C {Dollars per Time Period}
DELTA = 1.2 {This value must be > one so that the cost vs q curve is convex.}
I = .05 {Dollars per Dollar per Time Period}
MC = B*DELTA*Q^(DELTA-1)
N = 10
ΔMC = (MC - DELAY(MC,DT,MC-.000104))/DT {Dollars per Unit Mineral
Extracted}

17.5 S-Shaped Substitution Model

In the previous section, we dealt with substitution in an indirect way, assuming that its effect is a decrease of the price of the material extracted from the reserve and that the substitute is perfect in the sense that once we ran out of the material from the mine the demand can be satisfied by the substitute. In this section, we model more explicitly the substitution process under alternate assumptions. The advantage of this approach is that it is more amenable to empirical applications.

Assume that a metal is provided from both mining operations and firms that recycle discarded metal. The recycled metal could be used as a substitute for the metal provided by the mines. For example, recyclable iron and copper are close substitutes to these metals obtained from virgin ores. Assume further that all mining and recycling firms operate in competitive markets. The share of the recycled material is at first low but increases fast, for example, due to the improvement of collection and distribution channels and the refinements of recycling technology. After some time, the expansion of recycling begins to slow down. Distribution channels may now all be well established and technologies well developed.

The share F of the recycled material can be modeled to follow an S-shaped curve. A convenient and versatile functional form is the logistic equation

$$\frac{dF}{dt} = \text{GAMMA} * F * (1 - F), \tag{17.6}$$

with GAMMA as a constant determining the slope of the resulting S-shaped curve.

Assume for simplicity that the metal from the recycling firms is offered at the same price as that from the mines. This assumption may be justified by the fact that the recycled metal may not be bought if it is more expensive than the metal from the mine. If the recycled product is less expensive, recycling operations can realize an economic rent by raising their price to that of the product from the mine. Furthermore, let us assume that recycling cannot take place forever because recycled materials may ultimately become dispersed or decrease in quality through added impurities or changes in physical and chemical characteristics. We have metal "leakage" from the system. Hence, recycling will require that mines produce metal that can ultimately be recycled. Therefore, in this model recycling is not a "backstop" technology; that is, cannot indefinitely supply the metal.

The model assumes that mine owners maximize the cumulative present value of their profits CPVP, given an initial resource endowment and a fixed technology. The latter is expressed through the cost function. The Hamiltonian is

$$H = [P * QM - C]e^{-I * t} - \lambda * QM, \tag{17.7}$$

with QM as the materials extracted from the mine. These are sold on the market together with the substitute. Note that (17.7) is in essence the same as (16.6) in Sect. 16.1.

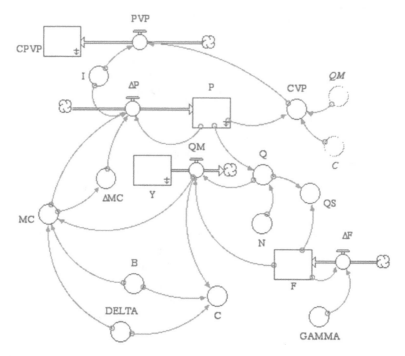

Fig. 17.6 Scarcity with substitution

The quantity QS of the substitute is defined by

$$QS = F * Q, \tag{17.8}$$

where Q is the demand met by one of N identical firms based on the linear demand curve of the previous chapters

$$Q = \frac{(10 - P)/0.00338}{N} \tag{17.9}$$

The STELLA model is shown in Fig. 17.6. Again, we need to find the initial price that maximizes CPVP, simultaneously driving reserves Y and the rate of extraction from the mine QM to 0 in the terminal time. The transversality conditions require (see Chap. 16) that $Y(T) = 0$ and so we seek the largest Po that just exhausts the resource. One can easily show that this Po, of all those meeting the resource exhaustion condition, also maximizes the CPVP.

When any of these firms dealing with scarcity begin their price raise they are acquiring increased risk that the higher price will produce a substitute. This risk is no doubt represented in their chosen discount rate. Change I to 12% and rerun the model.

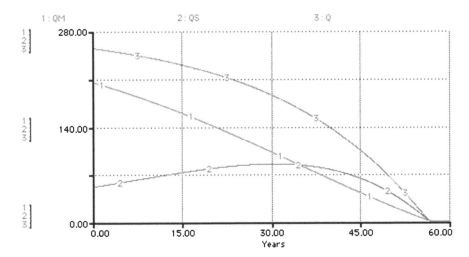

Fig. 17.7 Production quantities

Choose GAMMA $= 0.02$, an initial recycling fraction $F = 0.2$ and assume the same cost function as in the previous section. Run the model for $DT = 0.1$ and find the optimal initial price. Compare the results to those in the absence of recycling, that is, dF/d$t = 0$ and $F = 0$.

The graphs in Figs. 17.7–17.9 show that total demand for the metal drops. This is due to an increase in price in response to resource depletion in the mine. The quantity recycled is at first low but increases as the rate of recycling F increases. However, because mining output continuously decreases, the increase in F is not sufficient to maintain, or further increase, a temporarily achieved high level of recycling. Mining output and recycling drop to 0 by the terminal time.

Introduce a collection operation into the model by assuming that not all metal extracted from the mine is readily available for recycling; rather, it takes time for the metal to be discarded and enter the recycling process. Only when discarded can the material be collected and recycled. How does the average lifetime of the product, the time before it is discarded, influence the optimal price and the lifetime of the mine? Change the specification of the recycling fraction F such that F does not approach 1 for large values of t but some number $F < 1$.

Make the recycling fraction a function of cumulative extraction from the mine, rather than the rate of extraction. Set up the recycling process such that, once the mine is exhausted, recycling can supply the metal indefinitely; that is, model the development of a backstop technology.

Now that we have explored various ways of modeling substitution, let us return to the basic model of Chap. 16 and account for the effects of extraction history on cost of extraction. Particularly, we wish to model the fact that the more of

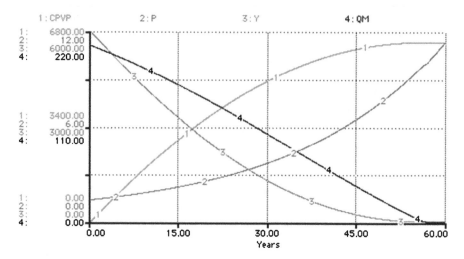

Fig. 17.8   Mining operation

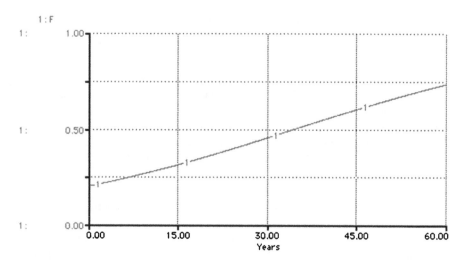

Fig. 17.9   Market share of recycled material

the resource already extracted, the higher is the extraction cost. We already have modeled in a simplified form the depletion effect on the cost of extraction in Chap. 3. Its inclusion in the models of optimal resource depletion is the topic of the following chapter.

17.6   S-Shaped Substitution Model Equations

CPVP(t) = CPVP(t - dt) + (PVP) * dt
INIT CPVP = 0{The cumulative present value of the profit }
INFLOWS:
PVP = CVP/(1+I)^TIME {Dollars per Time Period}
F(t) = F(t - dt) + (ΔF) * dt
INIT F = .2 {Tons per Ton}
INFLOWS:
ΔF = GAMMA*F*(1-F)
P(t) = P(t - dt) + (ΔP) * dt
INIT P = 1.378 {$/Ton}
INFLOWS:
ΔP = I*(P - MC) + ΔMC {Dollars per Ton per Time Period}
Y(t) = Y(t - dt) + (- QM) * dt
INIT Y = 60000/N {Tons}
OUTFLOWS:
QM = (1-F)*Q {Tons of Minerals from the Mine per Time Period}
B = .1
C = B*QM^DELTA {Dollars per Ton}
CVP = P*QM-C {Dollars per Time Period}
DELTA = 1.2   {This value must be > one so that the cost vs q curve is convex.}
GAMMA = .04
I = .04 {Dollars per Dollar per Time Period}
MC = B*DELTA*QM^(DELTA-1)
N = 10
Q = (10 - P)/.00338/N   {Tons of Minerals per Time Period}
QS = F*Q {Tons per Time Period}
ΔMC = (MC-DELAY(MC,DT,MC-.000104))/DT

Chapter 18
Competitive Scarcity with Cost Dependent on Production Rate and Resource Size

Truth is the conformation of appearance to reality.

Alfred North Witehead

18.1 Production Rate and Resource Size Effects

In the preceding models, we assumed a competitive firm extracting a finite resource that has a cost function that is dependent on the rate of extraction. In reality, however, cost of extraction increases not only with the rate of extraction but also with decreasing stock size. The smaller are the remaining reserves, the deeper they may lie or the less the mineral is concentrated in the mine. The model of this chapter incorporates that stock effect into the cost function (Fisher 1981). As a result of this stock effect, the scarcity rent rate can no longer rise at the rate of interest, and we surpass the intuitive model's capacity to accurately follow the optimal production path.

Assuming the cost function

$$C = A * Q^{\wedge}\text{ALPHA} * (Y/Yo/N)^{\wedge}\text{BETA}, \tag{18.1}$$

the Hamiltonian is now

$$H = [P * Q - C]\mathrm{e}^{-I*t} - \lambda * Q, \tag{18.2}$$

A save-disabled version of STELLA® and the computer models of this book are available at www.iseesystems.com/modelingeconomicsystems.

M. Ruth and B. Hannon, *Modeling Dynamic Economic Systems*,
Modeling Dynamic Systems, DOI 10.1007/978-1-4614-2209-9_18,
© Springer Science+Business Media, LLC 2012

leading to the following adjoint equation:

$$-\frac{\partial H}{\partial Y} = \frac{d\lambda}{dt} = \frac{\partial C}{\partial Y} * e^{-I*t}.$$
(18.3)

Equation (18.3) combined with (16.7) from Sect. 16.1

$$\frac{\partial H}{\partial Q} = \left[P - \frac{\partial C}{\partial Y} \right] * e^{-I*t} - \lambda = 0$$

gives the solution for ΔP:

$$\Delta P = I * (P - MC) + MCY + \Delta MC/DT.$$
(18.4)

In the presence of the stock effect on extraction cost, the discounted scarcity rent rate has to rise not only at the rate of interest but also must compensate for the change in extraction cost due to marginal changes in the reserve size, $\partial C/\partial Q$, which we denoted MCY to distinguish them from MC. To calculate MC and MCY in STELLA, we have chosen the analytic approach because here cost is a function of two variables. STELLA easily finds total derivatives but partial derivatives are more difficult and would complicate the model unnecessarily. The transversality conditions (see Chap. 16) indicate that λ $(t) \geq 0$ as $t \geq \infty$, as does Q, since by experimenting, we note that $Y(t)$ tends to a fixed positive value (about 62 units per firm) as time $\rightarrow \infty$. From this we can determine that the price in time period zero (P_0) is approximately 2.3, which produces this joint asymptotic result. The resulting model is shown in Fig. 18.1.

As in the earlier forms of the scarcity model, a trial and error process must be used to find the initial price that maximizes the cumulative present value of the profit. We must also set, as before, the initial value of the MC and MCY functions that make use of the DELAY built-in function. This is a tricky process. MC and MCY are not control variables, and therefore, their initial values must agree with the first MC and MCY values generated by the cost function. These values, of course, depend on the initial choice of the control variable, price. So we juggle the initial MC and MCY until we are close to the first model-generated value of MC. Fortunately, once we are close to matching the value, the ultimate result is not too sensitive to small changes in the initial values of MC and MCY. Then, using a trial and error procedure, we solved for the initial price that maximized CPVP.

The extraction schedule and reserve decline are shown in Fig. 18.2 alongside the corresponding CPVP and scarcity rent rate. Figure 18.3 shows the price response. Price and the present value of the profit rate are made 0 at the same time by adjusting the initial price. By use of a table in STELLA, we can show that the scarcity rent rises

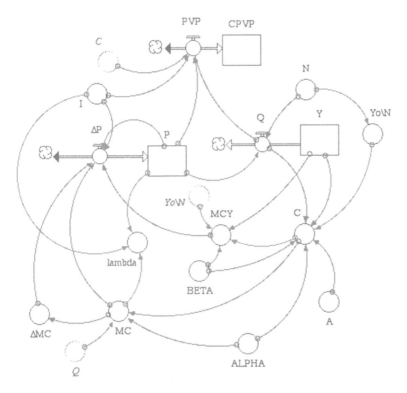

Fig. 18.1 Cost dependent on production rate and resource size

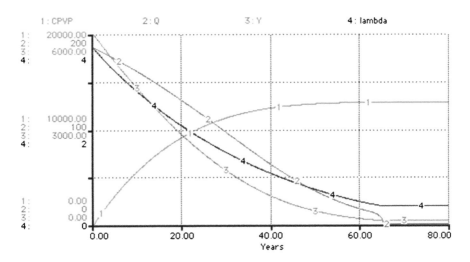

Fig. 18.2 Optimal extraction path, resource stock and value

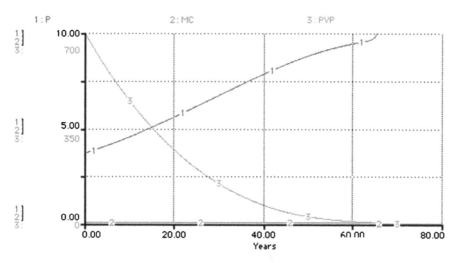

Fig. 18.3 Price, marginal cost and profit rate

more slowly than the exponential rate of the earlier models of this type. This is visually obvious toward the termination time of the resource extraction process. Note that we leave none of the mineral in the ground. The total profit is 89,400 and 6,300 resource units are left in the ground. Compare this result with that in Chap. 23.

Try varying the coefficients A, ALPHA, and BETA to find their effect on the lifetime of the reserve. These coefficients typically can be found from the historical analysis of data for this industry or firm. Each comes with an uncertainty attached, and one can explore the effects on the result by varying statistically estimated parameters within that range of uncertainty.

Finally, we should ask what the optimum extraction strategy would look like if we interrupted the optimal path, for example, at year 20, and reoptimized on the basis of the current reserve size and marginal costs.

Suppose that, in year 30, we were to find an additional 3,000 units of new resources. How would you handle this outcome? Show the entire extraction path in one graph in STELLA. In this model, increases in cumulative extraction from the resource stock lead to an increase in extraction cost. However, as the cost of extraction increases, the resource-extracting firms may find themselves compelled to improve their technologies to counteract the depletion effect on cost. Technology improvement would give them an edge on the market by being able to offer the resource at a lower price or by increasing the difference between the market price and their marginal cost of extraction. The following chapter discusses one form of technical change and its impacts on cost, price, and the lifetime of the mine.

18.2 Production Rate and Resource Size Effects
Model Equations

CPVP(t) = CPVP(t - dt) + (PVP) * dt
INIT CPVP = 0 {Dollars}
INFLOWS:
PVP = (P*Q - C)*EXP(-I*TIME) {Dollars per Time Period}
P(t) = P(t - dt) + (ΔP) * dt
INIT P = 3.7091990852
INFLOWS:
ΔP = ΔMC + I*(P - MC) + MCY   {Dollars per Ton of Minerals Time Period}
Y(t) = Y(t - dt) + (- Q) * dt
INIT Y = Yo\N {Tons of Mineral Reserve per firm}
OUTFLOWS:
Q = (10 - P)/.00338/N
A = .001 {Dollars per Ton of Minerals Extracted per Ton of Minerals in Mine per
Time Period}
ALPHA = 1.2
BETA = -.2
C = if Y > 0 then A*Q^ALPHA*(Y/Yo\N)^BETA else 0
I = .05 {Dollars per Dollar per Time Period}
lambda = (P-MC)*exp(-I*time)
MC = if Q > 0 then ALPHA*C/Q else 0 {Dollars per Ton of Minerals}
MCY = if Y > 0 then BETA*C/(Y/Yo\N) else 0
N = 10
Yo\N = 60000/N
ΔMC = DERIVN(MC,1) {Dollars per Ton of Minerals per Time Period}

Reference

Fisher AC (1981) Resource and environmental economics. Cambridge University Press, Cambridge

Chapter 19
Competitive Scarcity with Technical Change

> *Elegant intellects who despise the theory of quantity are but half developed.*
>
> A. N. Whitehead

19.1 Introduction

In contrast to the previous chapters on optimal resource extraction, let us more explicitly model changes in extraction technology. In this chapter, we model technical change that is endogenous to the system; that is, dependent on state variables of the system. Much of the technical change that takes place is a gradual improvement in production technologies that can be modeled as learning by doing: Increases in cumulative extraction will lead to increased experience in the mining operation. This increase in experience, in turn, translates into efficiency improvements and, thus, reductions in the average cost of the mineral extracted. Of course, these improvements cannot go on forever. If any of the inputs in the extractive operation is costly and essential, an assumption that is very likely to hold, then some minimum expenditures in the production process must be made that cannot be overcome by technical change. The laws of thermodynamics, for example, tell us that all real-world processes require a minimum of energy inputs (Ruth 1993, 1995). Furthermore, technical change itself may require materials or energy from those reserves whose scarcity it is supposed to alleviate.

You may recall that we have encountered simple versions of this learning-by-doing effect on cost in Chaps. 3 and 5. In each case, we have drawn cost curves as functions of cumulative production. When drawing these curves, we made sure that they declined at a decreasing rate and asymptotically approached a value

---

A save-disabled version of STELLA® and the computer models of this book are available at www.iseesystems.com/modelingeconomicsystems.

M. Ruth and B. Hannon, *Modeling Dynamic Economic Systems*,
Modeling Dynamic Systems, DOI 10.1007/978-1-4614-2209-9_19,
© Springer Science+Business Media, LLC 2012

significantly larger than 0. The latter implicitly captures minimum input requirements for the production process. Additionally, the model in Chap. 3 captured
material and energy consumption to sustain technical change.

19.2   A Basic Model of Competitive Scarcity
with Technical Change

Similar to the models of Chaps. 3 and 5, we introduce here technical change as a
function of cumulative extraction. Technical change will be fast early on, when
extraction is high and can make a relatively large contribution to cumulative
production, and will slow down over time, if and when the rates of extraction
decrease. Furthermore, no technical change can take place if none of the resource is
extracted. These assumptions are consistent with everyday experience. For example, the first few times you play a musical instrument you can improve a lot. The
longer you practice, however, the smaller the incremental improvements
become, and if you stop practicing, you stop improving.

For our mining operation, average cost of production is a measure of the
efficiency of production. A learning curve that shows decreases in average unit
cost AC of extraction with cumulative production Z is

$$AC = CM + G * Z^{\wedge}PHI, \tag{19.1}$$

with CM, G, and PHI as constants. In this model, average cost declines toward a
minimum level, CM. The parameters G and PHI determine the slope of the average
cost curve; that is, the speed at which learning can take place. PHI is negative so
that increases in cumulative past production Z decrease average cost.

The state variables in this model are Y and Z with $dY/dt = -Q$ and $dZ/dt = +Q$.
Thus, the Hamiltonian is

$$H = [P * Q - AC * Q]e^{-I^{*}t} - \lambda_1 * Q + \lambda_2 * Q, \tag{19.2}$$

which gives us the optimality and adjoint conditions, with one control and two state
variables,

$$\frac{\partial H}{\partial Q} = [P - AC] * e^{-I^{*}t} - (\lambda_1 - \lambda_2) = 0, \tag{19.3}$$

$$-\frac{\partial H}{\partial Y} = d\lambda_1/dt, \tag{19.4}$$

$$-\frac{\partial H}{\partial Z} = d\lambda_2/dt. \tag{19.5}$$

Equation (19.3) yields

$$\left(\overset{\bullet}{P}- \overset{\bullet}{AC}\right) * e^{-I^{*}t} - I^{*}(P - AC) * e^{-I^{*}t} = \lambda_1 - \lambda_2 \tag{19.6}$$

and because $Z = Y(t = 0) - Y + Zo$, Eq. (19.4) yields

$$\overset{\bullet}{\lambda}_1 = Q * G * \text{PHI} * Z^{(\text{PHI}-1)} * e^{-I^{*}t}. \tag{19.7}$$

Similarly, (19.5) yields

$$\overset{\bullet}{\lambda}_2 = Q * G * \text{PHI} * Z^{(\text{PHI}-1)} * e^{-I^{*}t}. \tag{19.8}$$

Thus, the optimality and adjoint equations combine to

$$P = I(P - AC) + AC - 2 * Q * G * \text{PHI} * Z^{(\text{PHI}-1)}, \tag{19.9}$$

which we use in our STELLA model, after correcting for the length of DT, to calculate the change in price

$$P = I(P - AC) + \frac{AC}{DT} - 2 * Q * G * \text{PHI} * Z^{(\text{PHI}-1)} \tag{19.10}$$

As in the previous chapters, we can find the optimal initial price through trial and error in successive runs of the STELLA model. An additional requirement for this model is that we specify an initial value for Z. The calculation of the cumulative present value of profits, price, total cost, and extraction are shown in Fig. 19.1.

Start again with an initial reserve size Y of 60,000/N tons, set $N = 10$, and use the same demand curve as in the previous section. Run the model at a DT = 1/64 with $G = 30$, PHI $= -0.5$, and CM $= 1$. You will find an optimum initial price of $2.371, resource exhaustion in approximately 41 years and a cumulative present value of profits CPVP = $6,350. An appropriate starting value for Z is 8,000. Average extraction costs drop from an initial value of $1.65/ton to approximately $1.33/ton. The results are shown in Figs. 19.2 and 19.3. The transversality conditions (see Chap. 16) are more difficult to apply here. First we know that lambda2(T) = 0, since lambda2(T)*Z(T) must be 0 and we know that Z(T) cannot be 0. Then from (19.3), we have at the terminal time: $(P-AC)e^{\wedge}(-I*T) = $ lambda1 (T) and since $P > AC$ always, lambda1(T) > 0. Therefore, $Y(T) = 0$, and the resource must be used up completely by the end of this process. Since no T was specified at $t = 0$, $H(T) = 0$ since $Q(T) = 0$.

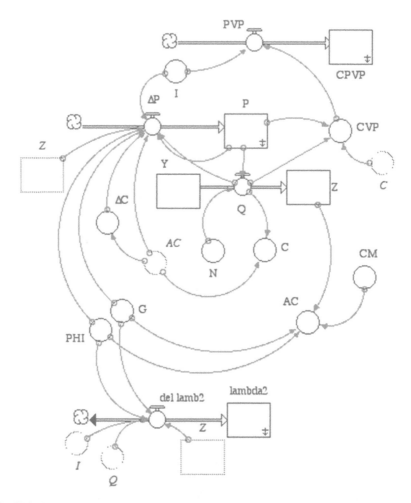

Fig. 19.1 Competitive scarcity with technical change

"Switch off" technical change, which resulted in the learning curve of Fig. 19.3. What are the effects on the optimal price, terminal time, and CPVP if there is no technical change? Make an educated guess before you run the model. Change the speed at which learning takes place and interpret your results.

Technical change is one way by which the price of nonrenewable resources may temporarily be held at a low level. Another way is through new discoveries. We assumed that those discoveries are exogenous to the model—we accidentally stumbled across a previously unknown field of oil or vein of ore. The following chapter takes up the topic of discoveries again but makes the more realistic assumption that a resource-extracting firm needs to actively engage in exploration efforts if it intends to find a new reserve.

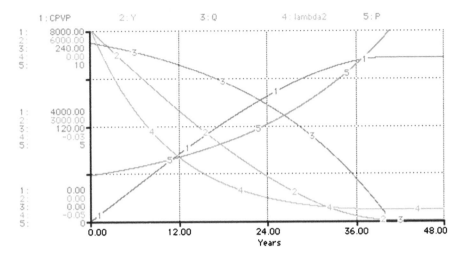

Fig. 19.2 Optimal extraction with technical change

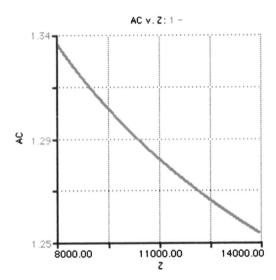

Fig. 19.3 Learning curve

19.3   Competitive Scarcity with Endogenous Technical Change Model Equations

INIT CPVP = 0 {Dollars}
INFLOWS:
PVP = CVP/(1+I)^TIME {Dollars per Time Period}
lambda2(t) = lambda2(t - dt) + (del_lamb2) * dt
INIT lambda2 = 0
INFLOWS:
del_lamb2 = Q*G*PHI*Z^(PHI-1)*EXP(-I*time)
P(t) = P(t - dt) + (ΔP) * dt
INIT P = 2.395   {Dollars per Ton}
INFLOWS:
ΔP = I*(P - AC) + ΔC -2*Q*G*PHI*Z^(PHI-1) {Dollars per Ton per Time Period}
Y(t) = Y(t - dt) + (- Q) * dt
INIT Y = 60000/N {Tons}
OUTFLOWS:
Q = (10 - P)/.00338/N {Tons per Time Period}
Z(t) = Z(t - dt) + (Q) * dt
INIT Z = 8000
INFLOWS:
Q = (10 - P)/.00338/N {Tons per Time Period}
AC = CM + G*Z^PHI {Dollars per Ton}
C = AC*Q {Dollars per Time Period}
CM = 1 {Dollars per Ton}
CVP = P*Q-C {Dollars per Year}
G = 30
I = .05 {Dollars per Dollar per Time Period}
N = 10
PHI = -.5
ΔC = DERIVN(AC,1) {Dollars per Ton per Time Period}

References

Ruth M (1993) Integrating economics, ecology and thermodynamics. Kluwer, Dortrecht
Ruth M (1995) Thermodynamic implications for natural resource extraction and technical change in U.S. copper mining. Environ Res Econ 6:187–206

Chapter 20
Competitive Scarcity with Exploration

*The generally accepted view is that markets are always
right—that is, market prices tend to discount future
developments accurately even when it is unclear what those
developments are. I start with the opposite point of view.
I believe that market prices are always wrong in the sense
that they present a biased view of the future*

George Soros, 1987

20.1 Scarcity and Exploration

In the previous models, we assumed the extractive process to be the only force
changing the reserve size. In reality, other forces play a role, too, in determining the
size of the reserve. Among those are exploration and discovery. The rate of
exploration E (e.g., the rate of feet drilled per year in search of oil) controls the
discovery rate (D), which, in turn, directly affects discovery cost (DC). Assume that
this relation between exploration and discovery rate is given by

$$D = 20^*E^\wedge \mathrm{BETA} \tag{20.1}$$

and that the relation between discovery cost and the rate of exploration is

$$DC = 3^*E^\wedge \mathrm{ALPHA} \tag{20.2}$$

The change of the exploration cost per unit of the resource found is called the
marginal discovery cost (MDC).

A save-disabled version of STELLA® and the computer models of this book are available at
www.iseesystems.com/modelingeconomicsystems.

M. Ruth and B. Hannon, *Modeling Dynamic Economic Systems*,
Modeling Dynamic Systems, DOI 10.1007/978-1-4614-2209-9_20,
© Springer Science+Business Media, LLC 2012

Here, we see an increase in our difficulty to find the correct numerical solution—now we have two control variables: exploration rate E and Q. We must find the optimum initial values for each of them.

Let us extend our intuitive understanding of the dynamics of resource depletion to see the effect of exploration and discovery on the optimal extraction path. If the MDC exceeds the current scarcity rent rate (the profit on extracting a unit of the resource now), exploration is proceeding too fast. Alternatively, if MDC is less than the current scarcity rent rate, the profit rate is too small—higher profits could be achieved if extraction were slowed down a bit. The resource owner is not replacing the last unit sold with a new resource at the breakeven rate.

This reasoning, however, is not always true. One can show that a wedge of cost emerges between MDC and the scarcity rent rate—the opportunity cost of finding the next unit of the resource. For our purposes we will ignore this opportunity cost. Therefore, we must devise an arrangement that will cause MDC to track the scarcity rent rate being generated in the model. We do this by controlling the exploration rate. In the model that follows the exploration rate is set so that the tracking process begins on the first year of resource extraction. Try raising the exploration rate and see what happens.

We can also extend the theory developed in the previous problem to account for exploration and discovery. The Hamiltonian capturing the discovery process (but not the opportunity cost of finding the next unit of the resource) is

$$H = [P^*Q - C - \text{DC}]e^{-I^*t} + (D - Q) \tag{20.3}$$

and the resulting optimality equations for the model are

$$\frac{\partial H}{\partial Q} = [P - \text{MC}]^* e^{-I^*t} - \lambda = 0, \tag{20.4}$$

$$\frac{\partial H}{\partial E} = -\frac{\partial \text{DC}^*}{\partial E} e^{-I^*t} + \lambda \frac{\partial D}{\partial E} = 0. \tag{20.5}$$

Equations (20.4) and (20.5) can be written as

$$\frac{\partial \text{DC}/\partial E^*}{\partial D/\partial E} e^{-I^*t} = \frac{\partial \text{DC}^*}{\partial D} e^{-I^*t} = \text{MDC}e^{-I^*t} = \lambda. \tag{20.6}$$

The adjoint equation in this problem is

$$-\frac{\partial H}{\partial Y} = \dot{\lambda} = 0. \tag{20.7}$$

These equations (20.4) and (20.6) reduce to

$$P = I(P - \text{MC}) + \text{MC}, \tag{20.8}$$

$$\text{MDC} = P - \text{MC}. \tag{20.9}$$

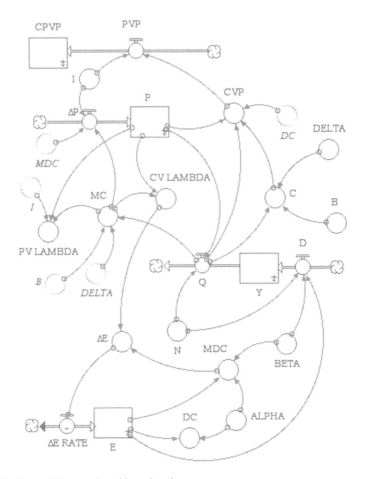

Fig. 20.1 Competitive scarcity with exploration

These results reaffirm our intuitive arguments that exploration will proceed with the marginal discovery cost equal to the scarcity rent rate.

Because $DC = 3^*E{\wedge}\text{ALPHA}$, $D = 20^*E{\wedge}\text{BETA}$, and with (20.6), we have in our model

$$P - \text{MC} = \text{MDC} = 3/20^*\text{ALPHA}/\text{BETA}^*E^{\wedge}(\text{ALPHA} - \text{BETA}). \qquad (20.10)$$

Since in this problem, infinity is the horizon, the applicable sufficiency conditions are: $\lim(t \to \infty)\lambda(t) \to 0$, $\lim(t \to \infty)\lambda(t)^*Y(t) = 0$, $\lim(t \to \infty)H(t) = 0$. We can see by inspection of (20.4) that $\lim(t \to \infty)\lambda(t) = 0$. The same is true for H due to $\lambda(t)$ and $e^{\wedge}(-I^*t)$ going to 0 in the limit. So there may or may not be a residual Y left in the ground in the long run.

In the main part of the model (Fig. 20.1), the initial price is adjusted to maximize the cumulative present value of profits. Because this adjustment will affect the

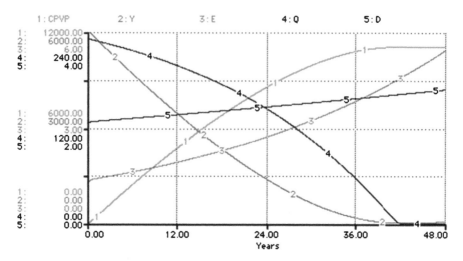

Fig. 20.2 Optimal resource extraction with exploration

scarcity rent rate, the initial value of the exploration rate may have to be adjusted. Since this latter adjustment will affect the discovery rate and hence the resource stock size, the price may have to be readjusted, although in our case the discovery rate is so small that it does not noticeably affect the solution. The adjustment process continues until the two criteria are met: tracking of the scarcity rent rate by MDC and the closure of the mining operation at the same time as the profit rate smoothly reaches 0. At this production–exploration rate combination, the cumulative present value of the profit is at a maximum (Fig. 20.2). You can demonstrate this point by moving the optimum set points for price and exploration rate and then calculating the CPVP for these cases. We can also show that the cumulative present value of the profit is less in the case of extraction with discoveries than extraction alone, even though the resource is exhausted here over a longer time. This resource owner would therefore not undertake any exploration. Even though the price to the consumer would be initially lower (higher later) and the resource would last longer, it reduces the owners' CPVP and they will not explore. Can you change the model to reverse this outcome?

In this problem, we have two control variables, price and exploration rate. We set the initial values of the control variables to produce the optimum production path. Try varying the initial level of the resource and see if you can readjust the model to its optimum level. Note how the delay functions are set to smooth out the initial Q and price curves.

From (20.9), we see that the discounted scarcity rent rate λ is also the discounted marginal discovery cost. So, we could examine a firm's marginal discovery cost and determine the level of its scarcity rate. This could tell us about the scarcity of the resource (Figs. 20.3 and 20.4). But imagine the firm learns as its finds new discoveries. It might learn where new discoveries lie, or it may learn that discovery

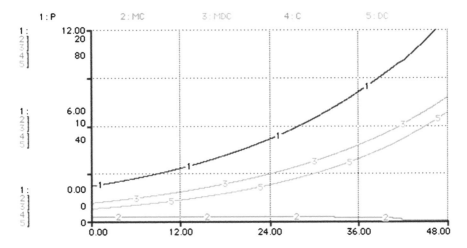

Fig. 20.3 Price and costs with exploration

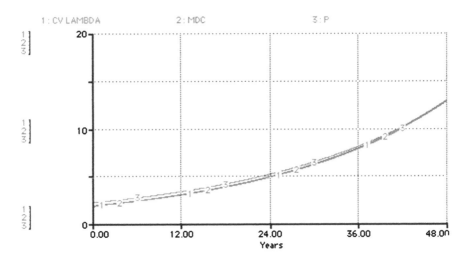

Fig. 20.4 Scarcity rent rate, marginal discovery cost and price with exploration

is becoming harder. This little insight makes the marginal discovery cost less valuable as a scarcity measure. There apparently are no simple measures of scarcity.

This model concludes our scenarios of optimal extraction of nonrenewable resources in a perfectly competitive market. The following chapters explore the case of a monopoly. We begin with a simple model that is analogous to the one developed in Chap. 16 and expand it in subsequent chapters to capture the effects of changes in the interest rate on optimal time paths and to include the impact of a declining reserve size on the cost of extraction.

20.2   Competitive Scarcity with Exploration Model Equations

CPVP(t) = CPVP(t - dt) + (PVP) * dt
INIT CPVP = 0 {The cumulative present value of the profit.
This is a maximum when the production rate and the reserves
have declined smoothly together to zero.}
INFLOWS:
PVP = CVP*EXP(-I*TIME)   {Dollars per Year}
E(t) = E(t - dt) + (ΔE_RATE) * dt
INIT E = 1.3 {This is the second control variable (price is the
other). It is controlled by the condition that the MDC should be
equal to the scarcity rent rate, an intutively reasonable guide.
Feet Drilled per Year.}
INFLOWS:
ΔE_RATE = GRAPH(ΔE)
(-0.2, 1.50), (-0.16, 1.39), (-0.12, 1.21), (-0.08, 0.96), (-0.04,
0.63), (1.39e-17, 0.27), (0.04, -0.255), (0.08, -0.75), (0.12, -
1.12), (0.16, -1.36), (0.2, -1.50)
P(t) = P(t - dt) + (ΔP) * dt
INIT P = 2.17
INFLOWS:
ΔP = I*(MDC) + DERIVN(MC,1) {Dollar per Unit of the Re-
source per Year}
Y(t) = Y(t - dt) + (D - Q) * dt
INIT Y = 60000/N {Units of the Resource}
INFLOWS:
D = 20*(E)^BETA/N {The given relation between the rate of
discovery of new units of resource and rate of exploration.
Units of the resource per Year.}
OUTFLOWS:
Q = (10 - P)/.00338/N {Units of the Resource per Year}
ALPHA = 1.6
B = .1
BETA = .2 {Units of the Resource per Feet Drilled}
C = B*Q^DELTA {This is the extraction cost and it is inde-
pendent of any stock size. The exploration cost is left out of
this cost because it would erroneously affect the marginal
cost. Its part of the total cost for the calculation of CVP. Dol-
lars per Year.}
CVP = P*Q-C-DC {Dollars per Year}
CV_LAMBDA = P - MC
DC = 3*E^ALPHA { The given exploration cost. It is not a
function of Q and therefore not included in the marginal cost
calculation, due to the nature of the MC calculation. Dollars
per Year.}
DELTA = 1.2 {This value must be > one so that the cost vs c
curve is convex.}
I = .04 {Dollars per Dollar per Year}
MC = B*DELTA*Q^(DELTA-1) {The marginal cost. It should
be a partial derivative but that is not possible on STELLA. Dol-
lars per Unit of the Resource.}
MDC = 3/20*ALPHA/BETA*E^(ALPHA-BETA) {The marginal
discovery cost: the dollar cost of a newly found unit of resource in the ground. Dollars per Unit of the Resource.}
N = 10
PV_LAMBDA = (P - MC)*EXP(-I*TIME)
ΔE = (MDC-CV_LAMBDA)/CV_LAMBDA {The variance of the
MDC from its target, the scarcity rent, normalized to control
the size of its variation. Feet Drilled per Year.}

Chapter 21
Monopoly Scarcity

> *Preposterous as it may seem at first blush, it is probably true*
> *that, even if all the timber in the US, or all the oil, or gas or*
> *anthracite, were owned by absolute monopoly, entirely free*
> *of public control, prices to consumers would be fixed lower*
> *than the long run interests of the public would justify*

> John Ise, 1925.

21.1 Nonrenewable Resource Extraction in a Monopoly

Let us return to the basic model of Chap. 16 but now assume that our firm has a
monopoly for the resource it extracts. As a monopolist, our firm could influence the
price via the amount of the mineral that it extracts from the mine. The extraction
rate will be chosen such that we maximize the cumulative present value of profits,
CPVP. The Hamiltonian in this model is

$$H = [P^*Q - C]^* \mathrm{e}^{-I^*t} - \lambda^* Q \qquad (21.1)$$

with

$$\frac{\partial H}{\partial Q} = \left[MR - \frac{\partial C}{\partial Q} \right]^* \mathrm{e}^{-I^*t} - \lambda = 0, \qquad (21.2)$$

$$-\frac{\partial H}{\partial Y} = \overset{\bullet}{\lambda}, \qquad (21.3)$$

as the optimality and adjoint equations. These equations are virtually the same as
those for the perfect competitors, with the exception that price is now a function

A save-disabled version of STELLA® and the computer models of this book are available at
www.iseesystems.com/modelingeconomicsystems.

M. Ruth and B. Hannon, *Modeling Dynamic Economic Systems*,
Modeling Dynamic Systems, DOI 10.1007/978-1-4614-2209-9_21,
© Springer Science+Business Media, LLC 2012

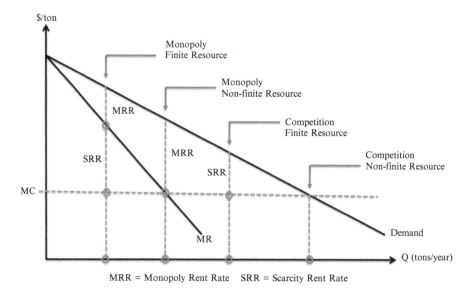

Fig. 21.1 Comparison of optimality conditions for competitive and monopolistic markets with finite and non-finite resources

of Q, and therefore, marginal revenues are no longer equal to the price. When you set up and run the model, remember to correct for the size of DT to calculate the change in MR over time, and ensure that the transversality conditions are fulfilled.

For the demand curve

$$P = 10 - 0.00338^*Q \tag{21.4}$$

marginal revenues are

$$MR = 10 - 0.00676^*Q \tag{21.5}$$

Solving the demand curve for Q allows the extraction rate to be determined from MR, which is a state variable in our model. Using the demand curve lets us find the price once Q has been determined. Price in the monopoly case is MR plus a monopoly rent rate (Fig. 21.1).

The marginal revenue curve for a straight-line demand curve is a straight line at twice the slope of the demand curve, as shown in (21.5) and the preceding sketch. Can you prove this? In the sketch, we show the four possibilities for market equilibrium for the competitive/monopoly, finite/nonfinite resource conditions. We have now covered all four of these conditions with our models (Fig. 21.1):

1. $P = MC$ for the pure competition condition with no finiteness in the inputs.
2. $P = MC + SRR$ (the scarcity rent rate) for the competition/finiteness condition.
3. $MR = MC$ and $P = MC + MRR$ (the monopoly rent rate) for the monopoly/nonfiniteness condition.
4. $MR = MC + SRR$ and $P = MC + SRR + MRR$ for the monopoly/finiteness condition.

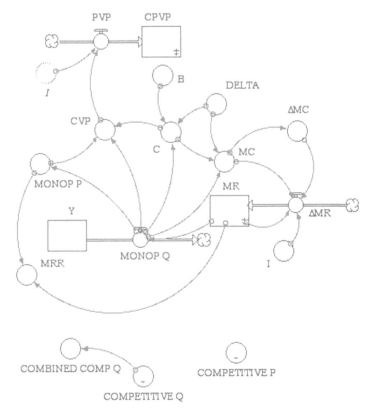

Fig. 21.2 Monopoly with finite resource

Note that the monopoly extracts both types of rent rates if it realizes that its input is finite. The rent rate is a possible scarcity indicator for the input resource. The problem lies in distinguishing between MRR and SRR, that is, in distinguishing the monopolist from the competitive firms that are using an increasingly scarce resource.

Before trying to find the rent rate and how it may have changed for a particular industry, one must first determine the degree of monopoly in the industry. One can assume away monopolistic possibility and then look for differences between *P* and *MC* for the industry. This approach, however, is hardly satisfying! As we have seen earlier, there exists an approximate surrogate for the SRR, the marginal discovery cost, i.e., the cost of finding the next unit of a resource. If that is not 0, then there probably exists some SRR and the resource is probably becoming scarce.

The model of a monopoly is solved here for a 60,000 unit finite resource (the same as Chap. 16) using trial and error values for the initial *MR* (Fig. 21.2). The *MR* is adjusted until the cumulative present value of profits, CPVP, is maximized. Use the Table Window for this effort.

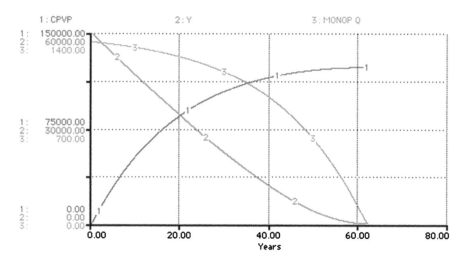

Fig. 21.3 Optimality conditions for nonrenewable resource extracting monopoly

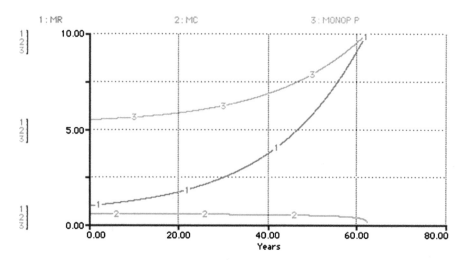

Fig. 21.4 Marginal revenue, marginal cost and price for nonrenewable resource extracting monopoly

The transversality conditions of Chap. 16, Sect. 16.1 apply directly here. $\lambda(T) > 0$ and therefore $Y(T) = 0$.

At a maximum CPVP, the monopolistic extraction rate and the resource stock have smoothly gone to 0 at the same time. All of the resource is used in this case (Fig. 21.3) and marginal revenue rises to meet price in the terminal period (Fig. 21.4).

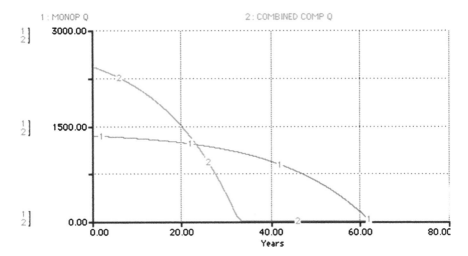

Fig. 21.5 Comparison of extraction rates for monopoly and competitive market

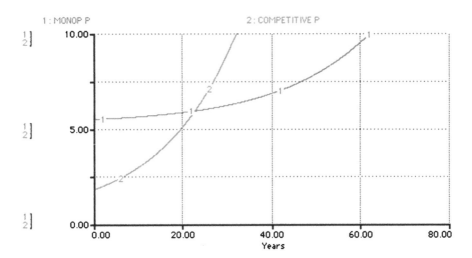

Fig. 21.6 Comparison of prices for monopoly and competitive market

We can now compare the competitive and the monopoly under scarcity. Begin the scarcity model of the ten firms in Chap. 16, each with 6,000 units of resource and the monopoly model of this chapter with the same amount, 60,000 units. The graphs in Figs. 21.5–21.6 depict the differences.

Note how much higher the price is here, in the initial and intermediate times, when compared with the competitive case (Chap. 16). Note also how much lower cumulative present value of profits is and how much longer the resource lasts (Figs. 21.5 and 21.6).

Is it always the case that monopolies will slow the rate of resource use? For the linear demand curve this is always true. But for a demand curve with a rising price–elasticity of demand, the monopolist will use the resource faster than the competitors! The price–elasticity of demand is

$$\frac{\partial Q}{\partial P} * \frac{P}{Q} \tag{21.6}$$

and an example of a demand curve with increasing price elasticity is

$$P = 120/Q^{\wedge}0.5 \tag{21.7}$$

Solve for the optimal extraction path under perfect competition and under monopolistic behavior, using this demand curve to show that the monopolist is now more "shortsighted" than the competitors. Because the monopoly profit is lower here than in the total competitive case, will monopolies be less likely to form in the finite resource markets than in regular markets? If a monopoly extends the life of a finite resource and accrues smaller cumulative present value profit, would you prefer that a monopoly controlled your society's finite resources? Before you jump to conclusions, recall the discussion we had in Chaps. 12 and 13 of the various effects a monopoly may have on the levels of pollution and on regional differences in output and employment. Recall also that discount rates, which play a primary role in determining the rate of extraction, are not constant over time and may differ among decision makers, as we argued in Chap. 9. The case of variable interest rates is modeled in the following chapter.

21.2 Monopolistic Scarcity

```
CPVP(t) = CPVP(t - dt) + (PVP) * dt
INIT CPVP = 0 {Dollars}
INFLOWS:
PVP = CVP*EXP(-I*TIME) {Dollars per Year}
MR(t) = MR(t - dt) + (ΔMR) * dt
INIT MR = .9448
INFLOWS:
ΔMR = I*(MR - MC) + ΔMC/DT {Dollars per Year}
Y(t) = Y(t - dt) + (- MONOP_Q) * dt
INIT Y = 60000 {Tons}
OUTFLOWS:
MONOP_Q = (10 - MR)/.00676 {Tons per Year}
B = .1 {Dollars per Ton}
C = B*MONOP_Q^DELTA {Dollars per Year}
COMBINED_COMP_Q = 10*COMPETITIVE_Q
```

COMPETITIVE_P = GRAPH(TIME)
(0.00, 1.81), (0.831, 1.88), (1.66, 1.96), (2.49, 2.04), (3.33,
2.13), (4.16, 2.22), (4.99, 2.31), (5.82, 2.41), (6.65, 2.51),
(7.48, 2.62), (8.31, 2.74), (9.14, 2.86), (9.98, 2.99), (10.8,
3.12), (11.6, 3.26), (12.5, 3.41), (13.3, 3.56), (14.1, 3.73),
(15.0, 3.90), (15.8, 4.08), (16.6, 4.27), (17.5, 4.46), (18.3,
4.67), (19.1, 4.89), (20.0, 5.12), (20.8, 5.36), (21.6, 5.62),
(22.4, 5.89), (23.3, 6.17), (24.1, 6.46), (24.9, 6.77), (25.8,
7.09), (26.6, 7.43), (27.4, 7.79), (28.3, 8.17), (29.1, 8.56),
(29.9, 8.97), (30.8, 9.39), (31.6, 9.82), (32.4, 10.2), (33.3,
10.7)
COMPETITIVE_Q = GRAPH(TIME)
(0.00, 242), (0.849, 240), (1.70, 238), (2.55, 235), (3.40, 233),
(4.25, 230), (5.10, 227), (5.95, 225), (6.80, 221), (7.65, 218),
(8.49, 215), (9.34, 211), (10.2, 208), (11.0, 204), (11.9, 199),
(12.7, 195), (13.6, 190), (14.4, 186), (15.3, 181), (16.1, 175),
(17.0, 170), (17.8, 164), (18.7, 158), (19.5, 151), (20.4, 144),
(21.2, 137), (22.1, 130), (22.9, 122), (23.8, 113), (24.6, 105),
(25.5, 95.6), (26.3, 86.0), (27.2, 75.9), (28.0, 65.3), (28.9,
54.3), (29.7, 42.7), (30.6, 30.6), (31.4, 18.1), (32.3, 5.40),
(33.1, 0.00)
CVP = MONOP_P*MONOP_Q-C {Dollars per Year}
DELTA = 1.2 {This value must be > one so that the C versu:
Q curve is convex.}
I = .05 {Dollars per Dollar per Year}
MC = DELTA*C/MONOP_Q {Dollars per Ton}
MONOP_P = 10 - .00338*MONOP_Q {Dollars per Ton}
MRR = MONOP_P-MR {Dollars per Ton}
ΔMC = MC-DELAY(MC,DT,MC+.0005) {Dollars per Ton}

Chapter 22
Monopoly Scarcity with Variable Interest Rate

Truth comes out of error more readily than out of confusion

Sir Francis Bacon

22.1 Introduction

Up to this point, we could solve many of the complex optimal depletion path problems with dynamic equations developed from our intuition of the behavior of a firm facing scarcity. We were able to generate the guiding differential equation by realizing that the firm would hold back on producing the resource, thus forcing up the value of the resource in the ground at the rate of interest. In this way, extraction was chosen such that a balance is reached between the profits to be made from selling the remaining resource and putting the money in the bank at the going interest rate or leaving enough of it in the ground so that its value would rise at the rate of interest. In so doing, the resource-extracting firm would maximize the cumulative present value of the profits from sales of the resource. With those revenues, the firm would generate a fund whose present value would supply the firm the same net income as it would have had if it were not dealing with a scarce resource.

The argument and calculation in those model sets were done under the assumption of a fixed interest rate. However, if the interest rate changes over time we face additional conditions that push us beyond our intuition. We have the possibilities of the firm expecting variable interest rates and possessing finite stocks whose extraction cost rises as the stock is depleted. Furthermore, we have the possible complexities of the firm wanting to optimize two paths at once: a production path and an exploration path.

A save-disabled version of STELLA® and the computer models of this book are available at www.iseesystems.com/modelingeconomicsystems.

M. Ruth and B. Hannon, *Modeling Dynamic Economic Systems*,
Modeling Dynamic Systems, DOI 10.1007/978-1-4614-2209-9_22,
© Springer Science+Business Media, LLC 2012

22.2   Monopoly, Scarcity, and Variable Interest Rates

How can we handle the complexity inherent in optimizing both a production and exploration path? We turn to a method that itself has limitations. It is a general procedure for reducing our optimality problem to one or more nonlinear differential equations. These equations are often not analytically tractable, but that is where numerical analysis and, in particular, STELLA comes in. We combine analytic procedure with numerical method to solve some of the more difficult problems in modern resource economics.

Let us return to the basic theory (Arrow and Kurz 1970). We assume that the finite resource problem is to maximize the cumulative present value of the profit, CPVP, over the lifetime of the resource. Here, we add the complexities of an interest rate that can vary with time and an extraction cost that is a function of both reserve size and rate of extraction:

$$\text{maximize CPVP} = \int_0^T (P^*Q - C(Q,Y))e^{-I(t)\,^*t}\mathrm{d}t. \tag{22.1}$$

From this equation and the constraint

$$\frac{\mathrm{d}Y}{\mathrm{d}t} = -Q \tag{22.2}$$

the Hamiltonian, H, can be set up very similar to the previous models as

$$H = (P^*Q - C(Q,Y))^* e^{-I(t)\,^*t} - \lambda^*Q, \tag{22.3}$$

where now the interest rate I is explicitly a function of time. This Hamiltonian yields as its optimality and adjoint equations:

$$\frac{\partial H}{\partial Q} = \left[MR - \frac{\partial C}{\partial Q}\right]^* e^{-I\,^*t} - \lambda = 0 \Rightarrow (MR - MC)e^{-I(t)\,^*t} = \lambda \tag{22.4}$$

and

$$-\frac{\partial H}{\partial Y} = \frac{\partial C}{\partial Y}^* e^{-I(t)\partial\,^*t} = MCY^* e^{-I(t)\,^*t} = \dot{\lambda}. \tag{22.5}$$

When (22.4) and (22.5) are combined to eliminate the costate variable λ, they yield the basic differential equation for the optimal resource extraction rate:

$$\frac{\mathrm{d}MR}{\mathrm{d}t} = \frac{\mathrm{d}MC}{\mathrm{d}t} + \left(I(t) + t^*\frac{\mathrm{d}I(t)}{\mathrm{d}t}\right)^*(MR - MC) + MCY. \tag{22.6}$$

If cost is not a function of Y and I is not a function of time, we find the intuitively derived equation that we have been using in the preceding problems,

$$\frac{\mathrm{d}MR}{\mathrm{d}t} = I^*(MR - MC) + \dot{M}C \tag{22.7}$$

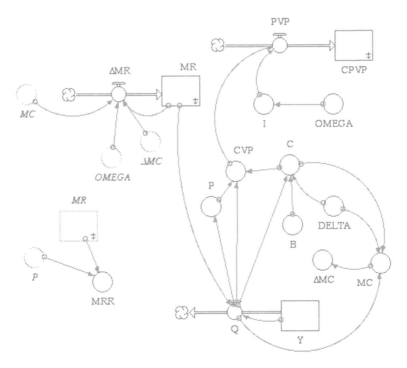

Fig. 22.1 Monopoly scarcity with variable interest rate

This simplified equation is the form most often talked about in the elementary literature of resource economics.

We will now solve the resource problem when the interest rate is not constant. In the next chapter we will take up the problem of a cost function that is dependent on the extraction rate and the size of the remaining resource.

Applying the transversality conditions of Chap. 16, we see from (22.5) that $\lambda(T)$ will not be zero for any finite T ($MCY = 0$) and therefore $Y(T) = 0$. We choose the largest Po that makes Y asymptotically zero, as this produces the largest CPVP for the constrained $Y(T)$.

Suppose that a monopoly, based on a finite resource, expects an interest rate that varies linearly with time:

$$I = I_0 + \Omega * t \tag{22.8}$$

The scarcity rent rate no longer rises at the same rate as the interest rate. Therefore, the interest rate in our model must be modified to change over time. Additionally, we must use (22.6) to capture the effect on MR of a dynamic interest rate. Then, the scarcity rent rate grows at $I + t * dI/dt = I_0 + 2 * t$ if the cost is not a function of the remaining resource size (Fig. 22.1).

In this problem, Ω is -0.0002 with an initial interest rate of 0.04. The results of the constant I runs are shown Figs. 22.2–22.4. The price starts slightly higher and the resource lasts many years longer. We could expect something like this as the

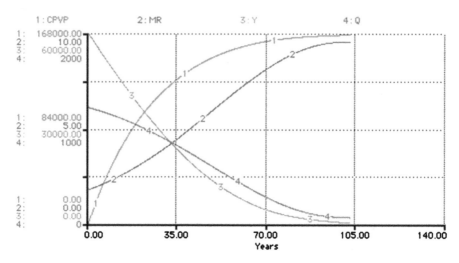

Fig. 22.2 Optimal extraction for monopoly facing a changing interest rate

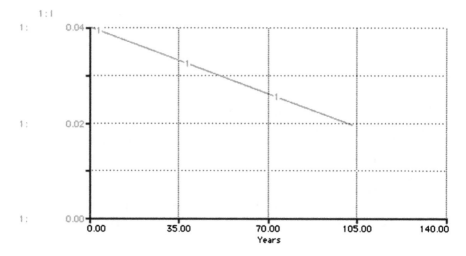

Fig. 22.3 Changing interest rate

scarcity rent rate must keep dropping at an increasing rate. Run the model with $\Omega = +0.0002$ and note the differences. Try to ramp the interest rate up from 4 to 6% in 20 years and hold it there for the rest of the run.

To derive the solution of this chapter, we assumed that there is no stock effect on the extraction cost. Of course, this is an assumption we may wish to eliminate, and we will do so in the following chapter.

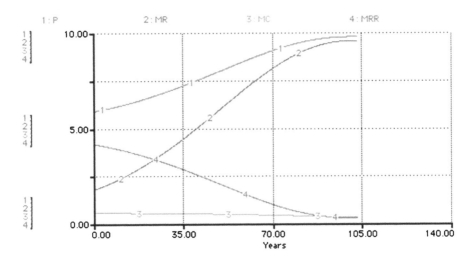

Fig. 22.4 Price, marginal revenue, marginal cost, and monopoly rent rate for optimal resource extraction under a changing interest rate

22.3   Monopoly Scarcity with Variable Interest Rate Model Equations

CPVP(t) = CPVP(t - dt) + (PVP) * dt
INIT CPVP = 0 {Dollars}
INFLOWS:
PVP = CVP*EXP(-I*TIME) {Dollars per Time Period}
MR(t) = MR(t - dt) + (ΔMR) * dt
INIT MR = 1.755 {Dollars per Ton}
INFLOWS:
ΔMR = (.04 + 2*OMEGA*TIME)*(MR - MC) + ΔMC/DT
{Dollars per Ton per Time Period}
Y(t) = Y(t - dt) + (- Q) * dt
INIT Y = 60000 {Tons}
OUTFLOWS:
Q = IF Y>=0 THEN (10 - MR)/.00676 ELSE 0   {Tons per
Year}
B = .1
C = B*Q^DELTA {Dollars per Year}
CVP = P*Q-C {Dollars per Year}
DELTA = 1.2 {This value must be > one so that the cost vs q
curve is convex.}
I = .04 + OMEGA*TIME {Dollars per Dollar per Time Period}
MC = DELTA*C/Q {Dollars per Ton}
MRR = P-MR {Dollars per Ton}
OMEGA = -.0002
P = 10 - .00338*Q {Dollars per Ton}
ΔMC = MC-DELAY(MC,DT,MC+.000062)

Reference

Arrow K, Kurz M (1970) Public investment, the rate of return and optimal fiscal policy. Johns
   Hopkins University Press, Baltimore, pp 26–57

Chapter 23
Monopoly Scarcity with Cost Dependent on Production Rate and Resource Size

Sometimes truth comes riding into history on the back of error.

Reinhold Niebuhr

23.1 Production Rate and Resource Size Influences on Monopolistic Resource Extraction

In this chapter, we return to the problem of extraction cost dependent on the rate of extraction and stock size. We laid out the analytical approach to that problem already in the previous chapter. However, let us not consider here a varying interest rate. You may combine the two issues, though, in one model. Before you do that, make sure to fully understand each problem by itself.

The cost function is designed to give about the same initial cost as in the previous chapter. It is specified as

$$C = A * Q^{\wedge}\text{ALPHA} * (Y/60{,}000)^{\wedge}\text{BETA}, \tag{23.1}$$

with $A = 0.1$, ALPHA $= 1.2$, and BETA $= 2$.

For the proper transversality conditions, see Chap. 18. The correct initial MR (1.29) produces $Y(T) = Q(T) = 0$ simultaneously is the one that finds $\lambda(T) = 0$ at the same time, uniquely satisfying the transversality conditions.

The addition of the reserve size effect to the cost function means also that the optimal path is affected by more than just the corresponding change in the MC. The interest rate must also be modified in the determination of the price by adding

A save-disabled version of STELLA® and the computer models of this book are available at www.iseesystems.com/modelingeconomicsystems.

M. Ruth and B. Hannon, *Modeling Dynamic Economic Systems*, Modeling Dynamic Systems, DOI 10.1007/978-1-4614-2209-9_23, © Springer Science+Business Media, LLC 2012

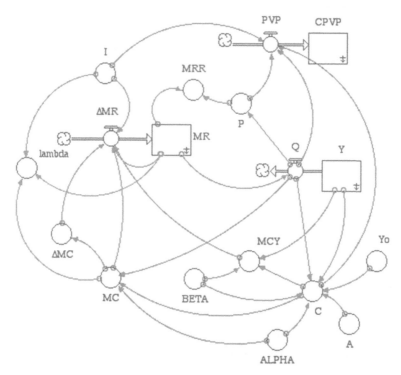

Fig. 23.1 Monopoly extraction with production rate and resource size affecting extraction cost

the partial derivative of cost with respect to the reserve size, divided by the scarcity rent rate (see the previous chapter for details). All this is done in the model of this chapter (Fig. 23.1). When you set up the model yourself, remember again to correct for the size of DT in the ΔMR calculation.

Introduce a production tax for the monopoly. Should this tax be considered a cost? Or should it just be subtracted from the price in the CPVP calculation? What do you think the mine owner will do about the tax? How will the resulting CPVP-maximizing extraction curve compare to the original one?

We can show analytically that the inclusion of the tax as a reduction in price leaves us with only an ambiguous solution: We cannot say which of many paths is best. But, if we include the tax as another cost, the solution is unique and reasonable. Try a tax of $0.2 per unit of Q, include it in the cost and reoptimize. You should find that the new optimal extraction curve begins at a lower rate and lasts a little longer than the path without the tax. The firm is pushing off in time the taxes that must be paid on the extraction. In this way, the present value of the taxes is less than had the firm not reoptimized! Proceed, as we did in Chap. 12, to include alternative forms of a tax on the monopolist. Set those taxes such that they generate the same *cumulative present value of tax revenues* as the production tax did and compare your results, with respect to the time it takes to exhaust the resource, and the optimal initial price (Figs. 23.2 and 23.3).

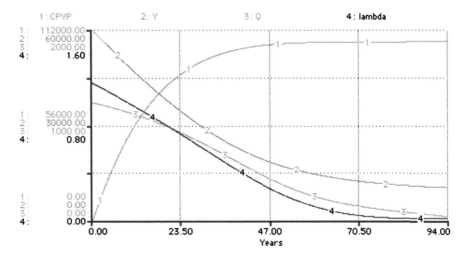

Fig. 23.2 Monopolistic optimum when extraction cost is dependent on production rate and resource size

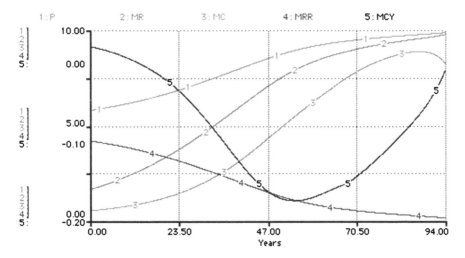

Fig. 23.3 Prices, marginal revenue, marginal costs, and monopoly rent rate

23.2   Production Rate and Resource Size Influences on Monopolistic Resource Extraction Model Equations

CPVP(t) = CPVP(t - dt) + (PVP) * dt
INIT CPVP = 0
INFLOWS:
PVP = (P*Q-C)*EXP(-I*TIME)
MR(t) = MR(t - dt) + (ΔMR) * dt
INIT MR = 1.6528
INFLOWS:
ΔMR = I*(MR-MC)+MCY+ ΔMC {Dollars per Ton per Year}
Y(t) = Y(t - dt) + (- Q) * dt
INIT Y = Yo
OUTFLOWS:
Q = (10-MR)/.00676 {This is the MR curve, solved for the output. Tons per Year}
A = .1
ALPHA = 1.2
BETA = -2
C = A*Q^ALPHA*(Y/Yo)^BETA {Dollars per Year}
I = .05
lambda = (MR-MC)*exp(-I*time)
MC = ALPHA*C/Q
MCY = BETA*C/Y
MRR = P-MR
P = 10-Q*.00338 {the demand curve.}
Yo = 60000
ΔMC = DERIVN(MC,1)

Part V
Modeling Optimal Use
of Renewable Resources

Chapter 24
Optimal Timber Harvest

> *I believe humanity made a serious mistake when our ancestors gave up the hunting and gathering life for agriculture and towns. That's when they invented the slave, the serf, the commissar, the master, the bureaucrat, the capitalists and the five-star General. Wasn't it farming that made a murderer of Cain? Nothing but trouble and grief ever since, with a few comforts thrown in now and then, like bourbon and ice cubes and free beer on the fourth of July, mainly to stretch out the misery.*
>
> Edward Abbey

24.1   Introduction

In this part of the book, we turn our attention to the harvest of renewable resources such as timber and fish. We concentrate on the optimal harvest by one of many identical firms that individually own part of that resource and operate in a competitive market. You may want to extend the following models to the case of monopolistic firms by using the insights of the previous chapters.

Assume you are an owner of a forestry operation that grows a single species of trees. Your operation is small in comparison to the rest of the tree growers in the country. A consequence of this assumption is that the price of trees is given and cannot be influenced by your production. Assume also for simplicity that the price of trees, measured in dollars per board feet of biomass, is constant and equal to 1. Figure 24.1 shows the changes in biomass Q over time. We deliberately misspelled time as TYME for reasons discussed later.

M. Ruth and B. Hannon, *Modeling Dynamic Economic Systems*, Modeling Dynamic Systems, DOI 10.1007/978-1-4614-2209-9_24, © Springer Science+Business Media, LLC 2012

Fig. 24.1 Growth of biomass

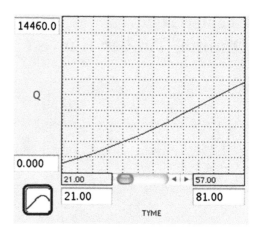

Fig. 24.2 Harvest cost as a function of biomass

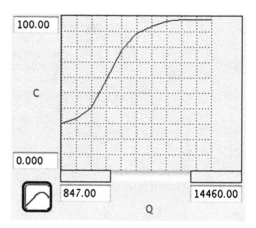

As the owner of the forestry operation you must make an initial investment of $100/ha in the plantation. Once you made this investment, for example, in the form of little saplings that you buy, trees grow and yield more biomass. This investment has to be made after each harvest. The cost of harvest (C) of trees depends on their biomass (Q) at the time of harvest. The relationship between harvest cost and biomass is given by Fig. 24.2.

No other costs occur during the growth phase of your trees. We also assume that soil fertility is constant. To maximize the cumulative present value of profits by your forest operation, you may want to find the optimum harvest time for a fixed interest rate and given the growth function of the renewable resource. Alternatively, you may want to fix the harvest time for successive runs and find the highest rate of return that can be achieved with that harvest time. The rate of return that can be achieved by managing the resource is called its *internal rate of return*. Let us employ the second method.

Alternative harvest times will yield different internal rates of return. Therefore, to find the maximum rate of return for alternative harvest times, we need to specify the harvest time, run the model, note the highest internal rate of return that is generated for that harvest time, and rerun the model for another harvest time. We should find that, for short rotation times of the forest, the maximum rate of return is low and that it increases with increasing rotation times, until a peak in the internal rate of return is reached. If the rotation times are increased further, the internal rates of return decrease again.

Let us recast this model in mathematical terms to facilitate its development in STELLA. As the tree plantation manager we face the initial investment of C_0. After t periods, the revenues from selling Q board feet of tree biomass at a price P is $P * Q$. At that period, we have to pay harvesting cost C that depends on Q and replant at the cost of C_0. This cycle of planting at cost C_0, timber growth, harvest, and new initial investment takes place into the distant future. Consequently, the present value of your profits is the stream of all the cost and revenues discounted at the interest rate I.

$$
\begin{aligned}
\text{PVP} &= -C_0 + (P * Q - C - C_0) * e^{-I*t} + (P * Q - C - C_0) * e^{-I*2*t} \\
&\quad + (P * Q - C - C_0) * e^{-I*3*t} + \cdots + (P * Q - C - C_0) * e^{-I*n*t} \\
&= (P * Q - C)(e^{-I*t} + e^{-I*2*t} + e^{-I*3*t} + \cdots + e^{-I*n*t}) \\
&\quad - C_0(1 + e^{-I*t} + e^{-I*2*t} + e^{-I*3*t} + \cdots + e^{-I*n*t}).
\end{aligned}
\tag{24.1}
$$

For simplicity of this analytical expression, let us define a new variable $a = e^{-I \times t}$, so we can rewrite the previous equation as

$$
\begin{aligned}
\text{PVP} &= (P * Q - C)(a + a^2 + a^3 + \cdots + a^n) - C_0(1 + a + a^2 + a^3 \\
&\quad + \cdots + a^n).
\end{aligned}
\tag{24.2}
$$

Our calculation of the cumulative present value of profits depends on the choice of the planning horizon. The longer our planning horizon, the more profits are accumulated. However, the further in the distance these profits occur, the lower is their contribution to the cumulative present value.

Ideally, we would like to determine the behavior of the forest operation over an infinite time horizon, assuming that the operation keeps growing trees over and over again. This requires infinite summation in the previous equation, which is, of course, quite a task to carry out. With a little mathematical trick, we can simplify matters considerably and get rid of this infinite sum of the values of a. For that purpose, let us define that sum as

$$
\begin{aligned}
s &= 1 + a + a^2 + a^3 + \cdots + a^n \\
\Rightarrow \quad as &= a + a^2 + a^3 + \cdots + a^n + a^{n+1} \\
\Rightarrow \quad (s - as) &= 1 - a^{n+1} \\
\Rightarrow \quad s(1 - a) &= 1 - a^{n+1}
\end{aligned}
\tag{24.3}
$$

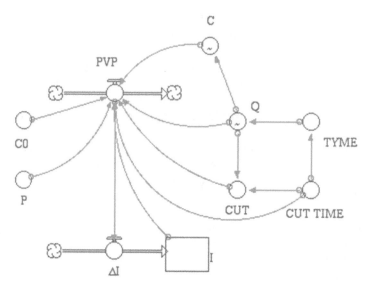

Fig. 24.3 Optimal tree cutting

and therefore

$$s = \frac{1 - a^{n+1}}{1 - a}. \tag{24.4}$$

Since $a = e^{-I*t}$, a^{n+1} will get very small for long time horizons, that is, for large values of n. We can, therefore, for all intents and purposes, neglect a^{n+1} in the numerator if we are interested in the behavior that is optimal over a very long time horizon and write

$$s = \frac{1}{1 - a}, \tag{24.5}$$

which we insert into the present value profit function to eliminate the sum of the values of a, leading to

$$\text{PVP} = (P * Q - C - C_0)/(1 - e^{-I*t}) - (P * Q - C) \tag{24.6}$$

This is the expression for the present value of profit rate function used in the STELLA model for the optimal harvest of trees (Fig. 24.3). The interest rate, I, is not fixed in the model. It is the internal rate of return that we vary to find the one that is largest, analogous to the way we have found the internal rate of return in Chap. 9.

The supporters of this method assert that one does not know the proper interest rate to use. The World Bank uses this method, but the economist Paul Samuelson soundly debunks it (Samuelson 1976). A full discussion of both methods can be found in Robinson (1972).

24.2   Optimal Harvest Time and Internal Rate of Return

Here, we set up the model such that the interest rate incrementally changes for a given CUT TIME; that is, the rotation time of the forest. To find the highest internal rate of return for that CUT TIME, we need to run the model over a time horizon that exceeds the growth of the first generation of trees.

We calculate a variable TYME that drives the change in biomass. Its values should start at 21, the first data point in the tree biomass graph just shown, increase over time, and be set back once the forest has been cut. Use the MODULO built-in function to generate a cyclical clock based on simulation time TIME:

$$TYME = MOD(TIME,\ CUT\ TIME - 20) + 21. \qquad (24.7)$$

We named this variable TYME so as to not confuse it with the built-in function TIME.

Start the model at some initial interest rate and increase that rate as long as the present value of profits is positive. Once the present value of profits reaches 0, you will have found the highest internal rate of return that corresponds to your chosen CUT TIME. Figure 24.4 shows the internal rates of return for rotation times of 24, 26, 28, 30, and 32 years.

The maximum internal rate of return is 10.6% for a cut time of 26 years. Figure 24.5 shows the internal rate of return, the present value profit rate and corresponding tree biomass for a CUT TIME of 26 years.

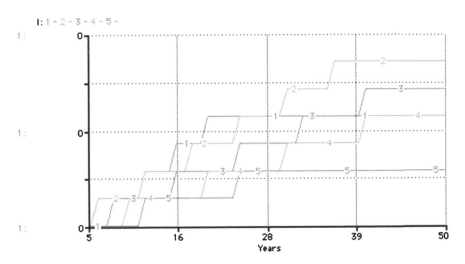

Fig. 24.4 Internal rates of return for alternative rotation times

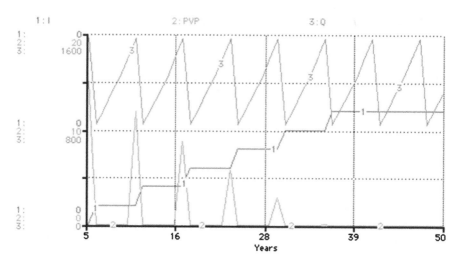

Fig. 24.5 Internal rate of return, profit, and biomass at the optimal rotation time

This method is rather controversial. The controversy stems from the concept of an internal rate of return. The word *internal* is the problem here. The return is not internal at all. The method determines the rate of return from forestland of a particular quality in a particular climate. It assumes that an infinite amount of such land is available so that profits from the timber operation can be plowed back into the operation. But can (and will) they? If not, then what is the rate of return of the alternate investment?

When trying to find answers to these questions, we immediately arrive at the problem of specifying the rate of return of the alternative investment to see if investing in tree farms is useful at all and, if it is, how long will it be before the tree growth slows to where it is more profitable to invest in the next most profitable alternative. Then we are back to the issue this method intended to circumvent.

The problem with this method is not that it is wrong. Rather, it is redundant and narrow. Only if there is more land of equal quality is this method of use, and then it will give only the same answer that can be found from the method of finding the cut time that maximizes the present value of the profit rate. To confirm this observation, set up the model to find the optimal cut time—rather than the internal rate of return for alternative cut times as we did here—and compare the results.

24.3 Internal Rate of Return Model Equations

I(t) = I(t - dt) + (ΔI) * dt
INIT I = .1
INFLOWS:
ΔI = IF PVP>0 THEN .001 ELSE 0
UNATTACHED:
PVP = IF CUT<>0 THEN ((P*Q - C-C0) /(1-EXP(-
I*CUT_TIME)) - (P*Q - C)) ELSE 0 {Dollars Per Time
Period}
C = GRAPH(Q)
(847, 30.0), (2208, 33.5), (3570, 40.5), (4931, 58.5), (6292,
77.5), (7654, 89.0), (9015, 93.5), (10376, 97.0), (11737,
98.0), (13099, 98.0), (14460, 98.0)
C0 = 100 {Dollars/hectare}
CUT = PULSE(Q,CUT_TIME-21,CUT_TIME-20)
CUT_TIME = 26
P = 1 {Dollars per board ft.}
Q = GRAPH(TYME)
(21.0, 847), (24.0, 1238), (27.0, 1730), (30.0, 2253), (33.0,
2848), (36.0, 3442), (39.0, 4087), (42.0, 4770), (45.0,
5555), (48.0, 6241), (51.0, 6955), (54.0, 7685), (57.0,
8418), (60.0, 9293), (63.0, 10032), (66.0, 10760), (69.0,
11478), (72.0, 12414), (75.0, 13154), (78.0, 13885), (81.0,
14460)
TYME = MOD(TIME,CUT_TIME-20) + 21

References

Robinson GA (1972) Forest resource economics. Ronald Press, New York
Samuelson PA (1976) Economics of forestry in an evolving society. Econ Inq 14:467–492

Chapter 25
Managing Open Access Resources

I know that history at all times draws strangest consequence from remotest cause.

T. S. Eliot, Murder in the Cathedral, Part I, 1935.

25.1 Tragedy of the Commons

In contrast to the model of the previous chapter, the same renewable resource may be exploited by many individual firms. If there is free access to the resource, many firms will compete with each other for what is left of the resource. The incentives for each individual firm to maintain a certain size of the resource can be quite low—possibly so low that too little of the renewable resource remains to ensure renewal. This is the tragedy of the commons; and it has been observed for many renewable resources, such as the grass on a commons jointly used by the members of a village, fisheries in open oceans, clean air, and clean water.

Let us model this process of unimpaired access to the commons and the resulting tragedy. Once we developed such a model, we can devise a procedure based on cooperation in the village that leads to the maximum number of sheep sustainable on the commons at a steady state.

Assume that the weather is constant and that the commons have a unit area with an initial grass stock of 250. This grass stock can reach a maximum 1,000 units. We model the growth of grass with a simple biological function. This function is shown in Fig. 25.1.

Assume that the only way the grass can be removed is through consumption by sheep. There are three age classes of sheep: 0–1-year olds, 1–2-year olds, and 3 or more year olds. Consumption of grass by sheep in each of the age classes is

---

A save-disabled version of STELLA® and the computer models of this book are available at www.iseesystems.com/modelingeconomicsystems.

M. Ruth and B. Hannon, *Modeling Dynamic Economic Systems*,
Modeling Dynamic Systems, DOI 10.1007/978-1-4614-2209-9_25,
© Springer Science+Business Media, LLC 2012

Fig. 25.1 Grass growth rate

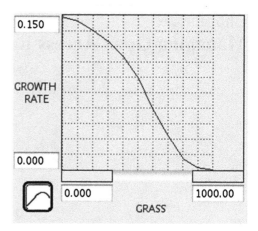

Fig. 25.2 Consumption
by the youngest age class

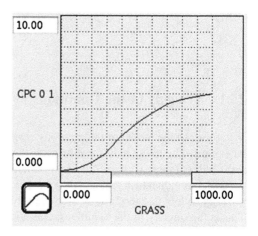

different, with young sheep consuming less than older sheep. Consumption per capita for each of the three age classes is denoted as CPC 0 1, CPC 1 2, and CPC 2 3, respectively, and defined by the graphical relationships with the stock of grass as shown in Figs. 25.2–25.4.

The part of the model that captures the growth and removal of grass is shown in Fig. 25.5. Total consumption of grass per time period is defined as the sum of grass consumption by sheep in each of the three cohorts:

$$\text{CONSUMPTION} = \text{CPC 0 1} * \text{AGE 0 1} + \text{CPC 1 2} * \text{AGE 1 2} + \text{CPC 2 3} * \text{AGE 2 3} \tag{25.1}$$

The death rates of sheep in each of the three age classes are assumed to depend on the per capita consumption of grass. If consumption is low, death rates are high. But even if per capita consumption is high, the death rate is not 0. Some sheep may die due to natural causes not related to their food supply. For each of the three age

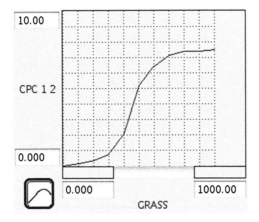

Fig. 25.3 Consumption by the middle age class

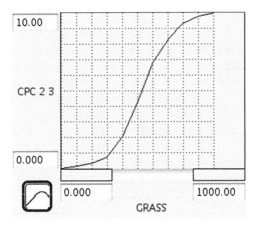

Fig. 25.4 Consumption by the oldest age class

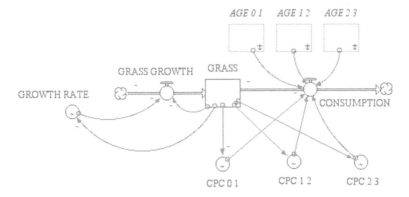

Fig. 25.5 Module for changes in the stock of grass

Fig. 25.6 Death rate

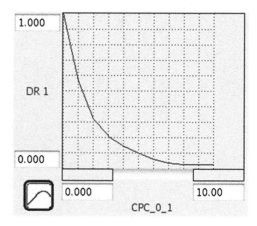

classes, the same graphical function for death rates is used. It is shown in Fig. 25.6 for the first age class.

Sheep get added to the commons by two processes. First, there is the natural increase in the sheep population due to births. Only sheep age 2 years and older can reproduce. The coefficient for the birth rate, expressed as a percentage of the individuals in the third cohort, is given as 12%. Second, sheep of all ages are added by the villagers to the commons. The additions occur at the beginning of each time period, and the sheep that do not die during the time period mature and move on to the next age cohort. Once sheep reach the last cohort, the villagers remove them at the end of the period; for example, to sell them on the market. The model for the three age cohorts on the commons is shown in Fig. 25.7.

Let us hold a set of assumptions on how sheep are added to the commons. First, we assume that sheep are added depending on the number of sheep already present on the commons but irrespective of the availability of the grass. Villagers following this strategy are very shortsighted. They take the increased number of sheep as a sign that the commons are in good condition. The more sheep are being added by other villagers, the more there is an incentive to compete with the other villagers and take advantage of the grass that is there. But, of course, the villagers are not totally blind to the conditions of the commons, and they may decide to not add any sheep of a particular age class if the number of deaths in that age class exceeds the number of sheep that mature into the next age class. The decision rules for the addition of sheep at each of the three age classes are given as

$$\text{ADD } 0\ 1 = \text{IF OUT } 0\ 1$$
$$\geq \text{DIE 1 THEN } A^*(\text{AGE } 0\ 1 + \text{AGE } 1\ 2 + \text{AGE } 2\ 3), \qquad (25.2)$$

$$\text{ADD } 1\ 2 = \text{IF OUT } 1\ 2$$
$$\geq \text{DIE 2 THEN } A^*(\text{AGE } 0\ 1 + \text{AGE } 1\ 2 + \text{AGE } 2\ 3), \qquad (25.3)$$

$$\text{ADD } 2\ 3 = \text{IF REMOVE } 2\ 3$$
$$\geq \text{DIE 3 THEN } A^*(\text{AGE } 0\ 1 + \text{AGE } 1\ 2 + \text{AGE } 2\ 3). \qquad (25.4)$$

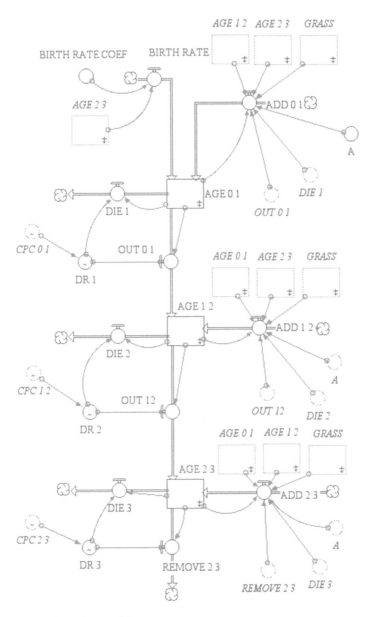

Fig. 25.7 Population cohort model

The parameter A is a measure of the competition or shortsightedness of the villagers and is measured in number of sheep added to a particular age cohort per total number of sheep on the commons. Let us set $A = 0.8$ and run the model. The results are fluctuating stocks of grass and sheep on the commons (Fig. 25.8). The cycles are vicious and perpetually repeated. Herd sizes collapse and stay at virtually 0 for a few periods, but temporarily recover due to restocking.

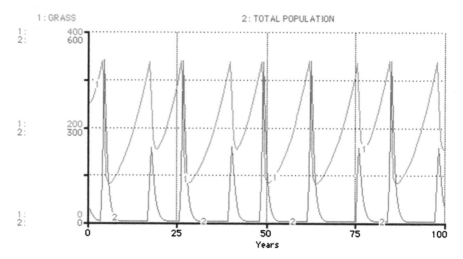

Fig. 25.8 Grass and sheep stock dynamics ($A = 0.8$)

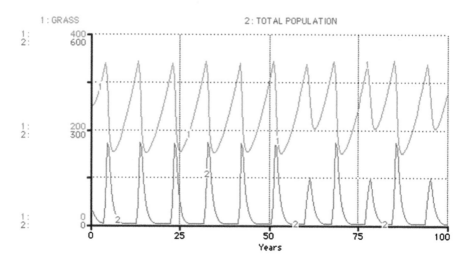

Fig. 25.9 Grass and sheep stock dynamics ($A = 0.4$)

In response to restocking too rapidly, too many sheep are on the commons. The sheep are decimating the grass and dying. After their death, the grass stock recovers and the commons receive new additions of sheep from the village.

The following three model runs are done for $A = 0.4$ (Fig. 25.9). Here, the additions of sheep to the commons are not as high as in the previous model run. Also, the uniformity of the cycles no longer holds, yet they are still strong, leading to collapse on the commons that is only temporarily overcome by restocking the herd.

We can easily calculate and plot the relative sizes of the three cohorts. Figure 25.10 shows that part of the model that provides a way of setting up such a calculation and a corresponding plot.

Fig. 25.10 Calculating relative cohort sizes

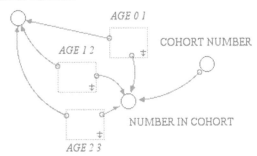

Fig. 25.11 Relative cohort sizes

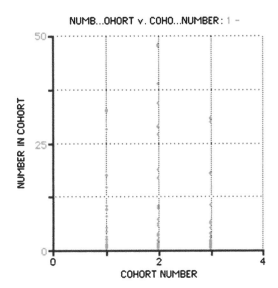

The cohort number is defined as

$$COHORT\ NUMBER = MOD(TIME/DT, 3) + 1. \tag{25.5}$$

It uses the built-in function to calculate the modulus and continuously counts from 1 to 3. The number of sheep in each of the three cohorts is then defined as

NUMBER IN COHORT = IF COHORT NUMBER = 1 THEN AGE 0 1
                  ELSE IF COHORT NUMBER = 2 THEN AGE 1 2
                  ELSE IF COHORT NUMBER = 3 THEN AGE 2 3
                  ELSE 0.

$$\tag{25.6}$$

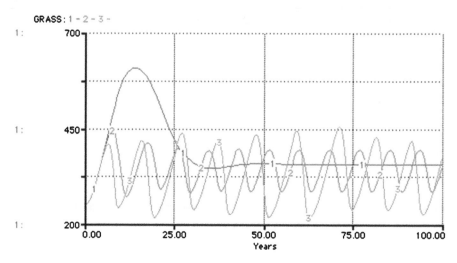

Fig. 25.12 Grass stock dynamics for alternate values of A

Plotting the number of sheep in each cohort against the cohort number yields the "bar chart" shown in Fig. 25.11. An alternative way to generate a bar chart—one, whose bars change as the model runs—is by clicking the "Bar" check box in the "Define Graph" field, which opens after double-clicking on a graph. Using our bar chart, which keeps track of values as the model runs, the results show a slightly underrepresented second age class, and the third age class marginally larger than the first age class. Can you explain why?

Assume that villagers are more conservative with their additions of sheep to the commons. Rerun the model for smaller values of A; then, increase A to model even more shortsighted villagers. The graphs in Figs. 25.12 and 25.13 show the results for $A = 0.2$, $A = 0.4$, $A = 0.6$.

An alternative strategy that the villagers may pursue in an attempt to avoid the tragedy of the commons is to consciously take the state of the commons into account in their decision-making process. Let us assume the following decision rules for adding sheep to the respective age classes:

$$\text{ADD } 01 = \text{IF OUT } 01 \geq \text{DIE 1 THEN } A^* \text{GRASS ELSE 0,} \qquad (25.7)$$

$$\text{ADD } 12 = \text{IF OUT } 12 \geq \text{DIE 2 THEN } A^* \text{GRASS ELSE 0,} \qquad (25.8)$$

$$\text{ADD } 23 = \text{IF REMOVE } 23 \geq \text{DIE 3 THEN } A^* \text{GRASS ELSE 0} \qquad (25.9)$$

and set $A = 0.15$. The fluctuations in the stock of sheep and grass still occur (Fig. 25.14), and the distribution of sheep is now skewed toward the younger age

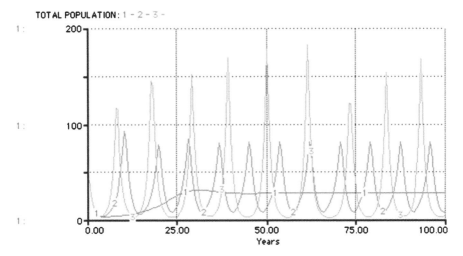

Fig. 25.13 Sheep stock dynamics for alternate values of *A*

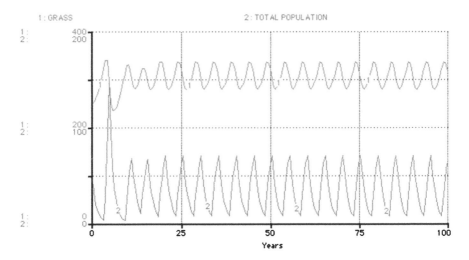

Fig. 25.14 Grass and sheep dynamics ($A = 0.15$)

classes (Fig. 25.15). Run the model for $A = 0.1$ and $A = 0.2$ to see the effect of the villagers' shortsightedness and competitiveness on the age distribution of sheep on the commons.

As we would expect, more conservative strategies dampen the fluctuations in the availability of grass and the size of the herd on the commons. However, are the villagers really better off if the fluctuations are less erratic? Assume that sheep can be sold on markets with given demand curves for each of the three age classes.

Fig. 25.15 Age distribution
($A = 0.15$)

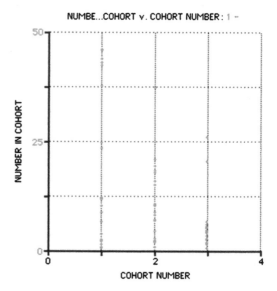

Assume that the price for younger sheep is lower than that for older ones. Villagers try to maximize the cumulative present value of their profits by selling sheep on these markets. For your model, assume the decision rules laid out in (25.3) through (25.5), and set $A = 0.2$, $A = 0.4$, $A = 0.6$ for consecutive runs. Then assume the decision rules of (25.7) through (25.9) and set $A = 0.1$, $A = 0.15$, $A = 0.2$ for consecutive runs. Assume a constant, positive discount rate and calculate for each of the models the cumulative present value of collective profits from selling the sheep. Interpret your results. What is the effect of combining the two decision rules on the cumulative present value of profits?

Can you find parameters A that yield a maximum steady-state sheep population on the commons for each of the two decision rules? How high is the cumulative present value of profits in each case, and how does this compare to the previous scenarios?

Assume that decisions to add sheep to the commons are not made immediately in response to the observed state of the commons. Rather, it takes villagers one full time period to decide how many sheep they should add. How does this time lag affect their cumulative present value of profits?

What are the impacts of a tax per number of sheep that are added each period on the fluctuations in the stocks of grass and sheep and on the cumulative present value of profits?

Now try making the weather cycle through the seasons, by having the grass senesce periodically and then regrow. Or, perhaps more appropriately, have the grass growth rate cycle between 0 and some maximum representing the seasonal effects. Assume $A = 0.2$ for your model runs. Does this setup make it harder to dampen the fluctuations in herd size?

25.2 Open Access Resources Model Equations

AGE_0_1(t) = AGE_0_1(t - dt) + (BIRTH_RATE + ADD_0_1 - OUT_0_1 - DIE_1
* dt
INIT AGE_0_1 = 10 {Individuals}
INFLOWS:
BIRTH_RATE = AGE_2_3*BIRTH_RATE_COEF {Individuals per Time Period}
ADD_0_1 = IF OUT_0_1 >= DIE_1 THEN A*(AGE_0_1+AGE_1_2+AGE_2_3) +
A*GRASS*0 ELSE 0 {Individuals per Time Period}
OUTFLOWS:
OUT_0_1 = (1-DR_1)*AGE_0_1 {Individuals per Time Period}
DIE_1 = DR_1*AGE_0_1 {Individuals per Time Period}
AGE_1_2(t) = AGE_1_2(t - dt) + (OUT_0_1 + ADD_1_2 - OUT_12 - DIE_2) * dt
INIT AGE_1_2 = 10 {Individuals}
INFLOWS:
OUT_0_1 = (1-DR_1)*AGE_0_1 {Individuals per Time Period}
ADD_1_2 = IF OUT_12 >= DIE_2 THEN A*(AGE_0_1+AGE_1_2+AGE_2_3) +
A*GRASS*0 ELSE 0 {Individuals per Time Period}
OUTFLOWS:
OUT_12 = (1-DR_2)*AGE_1_2 {Individuals per Time Period}
DIE_2 = DR_2*AGE_1_2 {Individuals per Time Period}
AGE_2_3(t) = AGE_2_3(t - dt) + (OUT_12 + ADD_2_3 - REMOVE_2_3 - DIE_3)
* dt
INIT AGE_2_3 = 20 {Individuals}
INFLOWS:
OUT_12 = (1-DR_2)*AGE_1_2 {Individuals per Time Period}
ADD_2_3 = IF REMOVE_2_3 >= DIE_3 THEN
A*(AGE_0_1+AGE_1_2+AGE_2_3) + A*GRASS*0 ELSE 0 {Individuals per
Time Period}
OUTFLOWS:
REMOVE_2_3 = (1-DR_3)*AGE_2_3 {Individuals per Time Period}
DIE_3 = DR_3*AGE_2_3 {Individuals per Time Period}
GRASS(t) = GRASS(t - dt) + (GRASS_GROWTH - CONSUMPTION) * dt
INIT GRASS = 250
INFLOWS:
GRASS_GROWTH = GRASS*GROWTH_RATE {Tons per Time Period}
OUTFLOWS:
CONSUMPTION = CPC_0_1*AGE_0_1 + CPC_1_2*AGE_1_2 +
CPC_2_3*AGE_2_3 {Tons per Time Period}
A = .15 {Individuals per Individuals}
BIRTH_RATE_COEF = .12 {Births per Individuals in Third Age Cohort}
COHORT_NUMBER = MOD(TIME/DT,3) + 1
CPC_0_1 = GRAPH(GRASS)
(0.00, 0.00), (100, 0.12), (200, 0.525), (300, 1.18), (400, 2.30), (500, 3.10), (600,
3.75), (700, 4.35), (800, 4.65), (900, 4.85), (1000, 5.00)
CPC_1_2 = GRAPH(GRASS)
(0.00, 0.00), (100, 0.12), (200, 0.33), (300, 0.75), (400, 2.05), (500, 5.15), (600,
6.45), (700, 7.15), (800, 7.40), (900, 7.40), (1000, 7.50)
CPC_2_3 = GRAPH(GRASS)
(0.00, 0.00), (100, 0.12), (200, 0.33), (300, 0.75), (400, 2.05), (500, 4.25), (600,
6.80), (700, 8.25), (800, 9.35), (900, 9.80), (1000, 10.0)
DR_1 = GRAPH(CPC_0_1)
(0.00, 1.00), (1.00, 0.56), (2.00, 0.32), (3.00, 0.21), (4.00, 0.145), (5.00, 0.1),
(6.00, 0.06), (7.00, 0.035), (8.00, 0.025), (9.00, 0.025), (10.0, 0.025)
DR_2 = GRAPH(CPC_1_2)

(0.00, 1.00), (1.00, 0.56), (2.00, 0.32), (3.00, 0.21), (4.00, 0.145), (5.00, 0.1),
(6.00, 0.06), (7.00, 0.035), (8.00, 0.025), (9.00, 0.025), (10.0, 0.025)
DR_3 = GRAPH(CPC_2_3)
(0.00, 1.00), (1.00, 0.56), (2.00, 0.32), (3.00, 0.21), (4.00, 0.145), (5.00, 0.1),
(6.00, 0.06), (7.00, 0.035), (8.00, 0.025), (9.00, 0.025), (10.0, 0.025)
GROWTH_RATE = GRAPH(GRASS)
(0.00, 0.148), (100, 0.145), (200, 0.136), (300, 0.124), (400, 0.11), (500, 0.0885),
(600, 0.0585), (700, 0.033), (800, 0.0112), (900, 0.00225), (1000, 0.00)
NUMBER_IN_COHORT = IF COHORT_NUMBER=1 THEN AGE_0_1 ELSE IF
COHORT_NUMBER=2 THEN AGE_1_2 ELSE IF COHORT_NUMBER=3 THEN
AGE_2_3 ELSE 0 {Individuals}
TOTAL_POPULATION = AGE_0_1+AGE_1_2+AGE_2_3 {Individuals}

Chapter 26
Optimal Catch from Fisheries

Nature must be considered as a whole if she is to be understood in detail.

G. Bunge, *Philosophical and Pathological Chemistry*, 1902

26.1   Optimal Fisheries Model

Issues of optimal use of renewable resources are typically complicated because resource owners need to take into considerations not only the size and quality of the resource as well as the technology and opportunity cost when extracting from a stock but also the growth dynamics characteristic of the biological populations they deal with. Additional problems are incurred in the case of free access, as we have seen in the context of common property resources. Management issues are still relatively tractable when dealing with resources that are easily monitored and "cultivated," as in the case of forests or sheep. Significant additional problems arise, however, when the resources are mobile, common property, and fluctuating in size considerably from year to year. This is the case for marine fisheries modeled in this and the following chapters.

To identify the optimal catch of fish we may use the approach chosen for the optimal extraction from a finite resource presented in the previous part of this book and augment the models by a natural growth process. Such a growth process can be captured in a way very similar to the elementary models of population growth we developed in Chap. 1.

The natural rate of growth G of the fish population is

$$G = \mathrm{GF}^* Y, \tag{26.1}$$

A save-disabled version of STELLA® and the computer models of this book are available at www.iseesystems.com/modelingeconomicsystems.

M. Ruth and B. Hannon, *Modeling Dynamic Economic Systems*,
Modeling Dynamic Systems, DOI 10.1007/978-1-4614-2209-9_26,
© Springer Science+Business Media, LLC 2012

Fig. 26.1 Relationship
between the growth factor
and population size

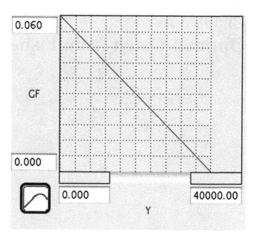

with GF the growth factor, which in turn, depends on the size of the population; that
is,

$$GF = GF(Y),$$ (26.2)

so

$$\frac{\partial G}{\partial Y} = GF + Y * \frac{\partial GF}{\partial Y}.$$ (26.3)

Here, we chose a maximum growth factor GF of 6% (when the reserve has
nearly vanished), on a straight line to a zero rate when the reserve is at 40,000
(Fig. 26.1).

As a result of the linear relationship between the population size Y and the
growth factor GF, the change in the growth factor with regard to a change in the
population size is just the slope of that line; that is,

$$\frac{\partial GF}{\partial Y} = -1.5 * E^{-6}.$$ (26.4)

The potential stock size increase must be captured in the analytical model used
to solve for the optimal extraction rate. The constraint equation for the state variable
Y is

$$\frac{dY}{dt} = G(Y) - Q.$$ (26.5)

The Hamiltonian for our model is

$$H = (P * Q - C(Q, Y)) * e^{-I^* t} + \lambda * (G(Y) - Q),$$ (26.6)

with cost dependent on the rate of extraction, or "individual catch", Q by an individual fishing operation and the size of the total fish population Y, and the growth rate is a function of stock size. As a result of the relationships among cost C, individual catch Q, and total stock Y, average cost AC and marginal cost MC all depend on an *individual's* Q and total Y. But Y is augmented by the natural growth of the stock Y and is depleted by the *total* catch, Q TOTAL, as determined from the demand curve and price P.

The optimality and adjoint equations for our model are

$$\frac{\partial H}{\partial Q} = (P - \text{MC})\, e^{-I^* t} - \lambda, \tag{26.7}$$

$$-\frac{\partial H}{\partial Y} = \dot{\lambda}, \tag{26.8}$$

which reduce to

$$\dot{P} = \left[I - \frac{\partial G}{\partial Y}\right] * (P - \text{MC}) + \dot{\text{MC}} - \text{MCY}. \tag{26.9}$$

Thus, very similar to the models of nonrenewable resources with discoveries, the change in the stock size requires an adjustment of the interest rate to capture the growth effect of the stock. In our model, this growth effect is reflected by the growth fraction itself. The corresponding transversality conditions must be fulfilled.

The STELLA model is shown in Fig. 26.2. It incorporates a module to tax fishermen, and thus increase the cost of fishing. Let us first set this tax-induced cost to 0. We return to it later.

This model is much trickier to run than earlier ones (Figs. 26.3 and 26.4). We must monitor the cumulative present value of the profit, CPVP, carefully to find the absolute maximum, but the result is very sensitive to the initial price. One can reason that the closer we are finally to a steady state, the larger the CPVP will be. This seems to be true, but you try it.

We have run the problem out to 100 years. It seems close to the steady state and the CPVP keeps rising ever more slowly. Try to lower the maximum growth rate in the graphed function to a rate of 0.035. Be sure to change the slope constant in the translation variable called ΔP. Can you conclude that the best result is to extinguish this fishery? Why or why not? Recall the problem of the tree cut. Would the hybridization of these fish to a faster growth rate tend to keep them from being destroyed, given the current fishing technology? Try a fisherman's discount rate of 20% find the final stock level. What resources do the fish depend on? How can we model the rise and fall of a fishery if we do not know the vagaries of the ecosystem on which the fishermen ultimately depend?

The graph in Fig. 26.5 shows the rise of the growth rate of fish as the stock is depleted. The curve has risen to a peak and then declined. Biologists refer to this

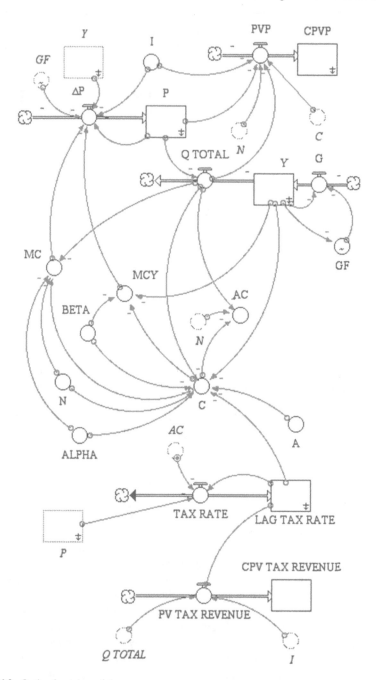

Fig. 26.2 Optimal catch model

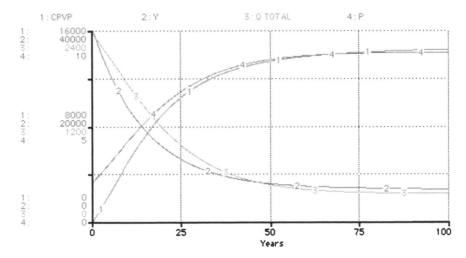

Fig. 26.3 Optimal catch, fish stock, price, and profit

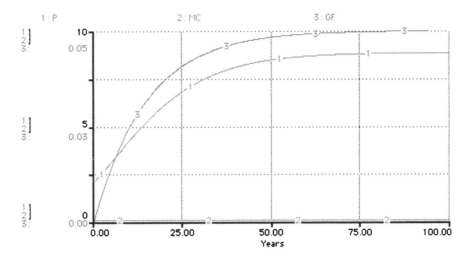

Fig. 26.4 Price, tax rate, marginal cost, and growth fraction for the optimal fishery

peak as the maximum sustainable yield (MSY) and recommend it as the operating point for fisheries. It is the most economically appropriate point when the interest rate and stock effects are ignored and environmental conditions remain constant. In this example, following the economic solution, the fishermen drive the stock of fish well below the MSY. As we see in the following, this can also be the case if there is some government intervention. What interest rate would be required to reach the MSY?

The idea of the fishery brings up again the potential for the "tragedy of the commons." The tragedy occurs when competitors ignore the finite potential of the

Fig. 26.5 Growth rate

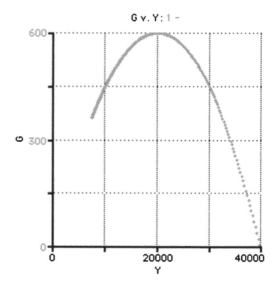

resource and collectively ignore the scarcity rent. In the world of real (not optimizing) fishing, the fishermen set their average cost equal to the price and compete away all profits. With price equal to average cost, the rate of use is higher and the stock is lower and more susceptible to extinction.

Most economists would favor a tax on the fish caught as a remedy to the tragedy. It would cover two items in order to drive the condition $P = AC$ governing fishermen's decisions market into a $P = MC +$ scarcity rent rate decision: One part, which we can calculate, would be equal to the scarcity rent rate; an added part would cover the difference between MC and AC. The tax is, therefore, equal to the price less the average cost in our optimal result.

Our preceding optimization shows the results when the fishermen follow their collective best interests and collect the scarcity rent. Under this condition, the initial price is $2.061525. But our claim is that they really would not do this for long. One or more of them will lower his price and sell all that he can catch. This price war would continue until the profit was entirely gone ($P = AC$). Unfortunately, the fish reserve would then be gone, too. To ensure that the reserve will be optimized, we raise a production tax on the fishermen whose rate is $P - AC$. This tax is treated as an additional cost, and the problem is reoptimized (Figs. 26.6 and 26.7). Now the initial price is $2.157 and the final steady state extraction rate and reserve size are about the same as before, but the cumulative present value of the private profit is lower. Additionally, the government can raise a significant tax revenue, shown in the model in its present value, PV TAX REVENUE, and as a cumulative present value CPV TAX REVENUE ($18,027 in $t = 100$).

Note well the extreme sensitivity of the extraction and reserve paths to the initial value of any of the stocks and the optimal starting price. The steady state paths for

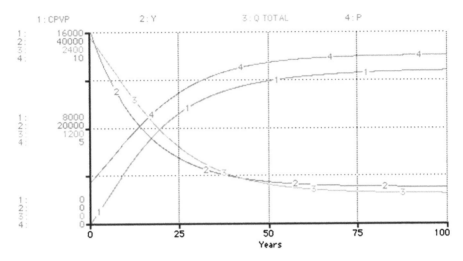

Fig. 26.6 Optimal catch, fish stock, price, and profit with the production tax as a cost

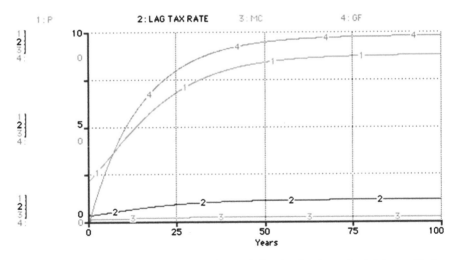

Fig. 26.7 Price, tax rate, marginal cost, and growth fraction for the optimal fishery with the production tax as a cost

any particular model, however, show very little difference. The fishermen reduce the extraction rate earlier in the period when the tax is applied to reduce its effect on the present value of the profit (which is now 0). So, with the government collecting this tax, the fishermen are operating with the profit level they would have sought on their own (0) as profit optimizers. With the tax, the steady state is reached rather than a collapse of the fishery. The results of the model with the inclusion of the tax follow.

Unfortunately, most fishery managers will try to restrict the fishing seasons and limit the fishing technology rather than implement a tax. The latter strategy will idle workers in the industry and cause the expansion of the technology into new, more effective, temporarily legal arenas. The tax is a far better way but an anathema to the independent-minded fisherman.

With the tax, government income increases and the fishery is perpetuated. The problem boils down to one of perception on the part of the firm owners and personal vs. governmental discount rates. We have alluded to this problem already in Chap. 9. Do the firm owners trust the government to accurately calculate market demand and their costs over the next 75 years? Perhaps the firm owners have a higher discount rate than the government. At the moment, the fishers would not care if the fishery were demolished in the next 10 years, but the government, with its supposed longer range view might want the fishery to last forever at some level and might want to avoid the unstable swings in the fishery size over the century. It is a classic question of government intervention. The fishermen may well realize that they should be taking a scarcity rent but are powerless to do so on their own. They may well understand that the unbridled competitive spirit will destroy the fishery in a short time but they can do nothing but adopt cost-cutting technology and ultimately sell at the average cost.

Under the tax program, cost-cutting technical improvements are also the only way that individual profits can be increased, so the tax spurs the use of new technology. The government is inclined to stay attentive because reduced average cost will reduce the selling price (AC + tax), and so the tax must be pushed up to compensate for average cost decreases in the industry in order to protect the fishery. The very instincts that cause the firm owners to sell at average cost will keep them adopting lower cost technologies under the government's plan: They assume that the industry as a whole is changing less rapidly than they are and so the industry price will fall less rapidly than they can drive down their own average costs. So, they assume, their own profit will rise faster than that of the industry, allowing them to expand their own operation's size faster than that of their competitors and thus they can achieve higher shares of the diminishing size of the fish population. It is a plan that will operate well if it could only get started. Thus, the main point of this discussion is that efficient intervention in the "commons" problem can save the industry from self-destruction and maintain competition with technological improvement. Are there non-market-based intervention mechanisms that may be beneficial to both the fish population and the fishing industry? Chapter 28 explores such an option—fish reserves. Before we discuss the introduction of reserves, let us develop a simple version of this model in Chap. 27 and investigate its behavior. In both of these chapters, we make extensive use of the experimental approach to modeling economic decisions that we have called for throughout this book.

26.2   Optimal Catch Model Equations

CPVP(t) = CPVP(t - dt) + (PVP) * dt
INIT CPVP = 0 {Dollars}
INFLOWS:
PVP = (P*Q_TOTAL-C)*EXP(-I*TIME)/N {Dollars per Year}
CPV_TAX_REVENUE(t) = CPV_TAX_REVENUE(t - dt) +
(PV_TAX_REVENUE) * dt
INIT CPV_TAX_REVENUE = 0 {Dollars}
INFLOWS:
PV_TAX_REVENUE = Q_TOTAL*LAG_TAX_RATE*EXP(-
I*TIME) {Dollars per Year}
LAG_TAX_RATE(t) = LAG_TAX_RATE(t - dt) + (TAX_RATE)
* dt
INIT LAG_TAX_RATE = .3
DOCUMENT: This stock is needed to avoid STELLA's circu-
larity check which is based on the drawing geometry rather
than the equation.
INFLOWS:
TAX_RATE = P - AC - LAG_TAX_RATE/DT {Dollars per In-
dividual}
P(t) = P(t - dt) + (ΔP) * dt
INIT P = 2.061525   {2.157 Dollars per Individual}
INFLOWS:
ΔP = DERIVN(MC,1) + (I+MCY/(P-MC) - GF + Y*1.5E-6)*(P-
MC) {Dollars per Individual per Year}
Y(t) = Y(t - dt) + (G - Q_TOTAL) * dt
INIT Y = 40000 {Individuals}
INFLOWS:
G = Y*GF   {Individuals per Year}
OUTFLOWS:
Q_TOTAL = (10-P)/.00338 {Individuals per Year}
A = .300
AC = C/Q_TOTAL/N {Dollars per Individual}
ALPHA = 1.6
BETA = -.3
C = A*(Q_TOTAL/N)^ALPHA*Y^BETA +
LAG_TAX_RATE*Q_TOTAL*0 {Dollars}
GF = GRAPH(Y)
(0.00, 0.06), (4000, 0.054), (8000, 0.048), (12000, 0.042),
(16000, 0.036), (20000, 0.03), (24000, 0.024), (28000, 0.018),
(32000, 0.012), (36000, 0.006), (40000, 0.00)
I = .04 {Dollars per Dollar per Year}
MC = ALPHA*C/Q_TOTAL/N {Dollars per Individual}
MCY = BETA*C/Y {Dollars per Individual}
N = 10

Chapter 27
Predator–Prey Models of Fisheries

And it ought to be remembered that there is nothing more difficult to take in hand, more perilous to conduct, or more uncertain in its success, than to take the lead in the introduction of a new order of things. Because the innovator has for enemies all those who have done well under the old conditions, and lukewarm defenders in those who may do well under the new. This coolness arises partly from fear of the opponents, who have the laws on their side, and partly from the incredulity of men, who do not readily believe in new things until they have had a long experience of them. Thus it happens that whenever those who are hostile have the opportunity to attack they do it like partisans, whilst the others defend lukewarmly.

N. Machiavelli, *The Prince*, 1513.

27.1 Basic Fisheries Model

In the previous chapter, we extended the model of optimal extraction from nonrenewable resources to enable renewal of the resource and then identified optimal harvest from a fish population. In this chapter, we choose an alternative approach to modeling fisheries. Methods of modern marine fisheries management virtually are no different from ancient hunter–gatherer strategies, in that the "prey" has to be first spotted, then caught or collected, and hauled to the home port. We therefore use predator–prey models to simulate the interactions between the economic and biological systems (Ruth 1995). The fish are the prey, and boats of fishermen correspond to the predators.

A save-disabled version of STELLA® and the computer models of this book are available at www.iseesystems.com/modelingeconomicsystems.

M. Ruth and B. Hannon, *Modeling Dynamic Economic Systems*, Modeling Dynamic Systems, DOI 10.1007/978-1-4614-2209-9_27, © Springer Science+Business Media, LLC 2012

The resulting interactions between the economic and biological systems may be simulated with well-known predator–prey models. Predator–prey models are frequently used in ecology and have been extensively analyzed. For example, Johan Swart (1990) elegantly presented the richness of even the simplest predator–prey models, showing damped and explosive oscillations as well as stable limit cycles. We encounter those features in the models developed here. The arguments presented can easily be generalized to other mobile renewable resources.

Much of the real-world fluctuations in the size of fish populations are caused by unpredictable environmental fluctuations that are poorly understood. Reference to those fluctuations is used to justify deviations of model predictions from actual population changes or to stress similarities between seemingly erratic model predictions and actual population changes (Wilson et al. 1990).

We contend that fish population dynamics may as well be attributed to economic factors that play an increasing role in the "predator–prey relationship." Many economic forces are poorly understood and difficult to forecast. Furthermore, differences between the model result and actual system behavior are significantly dependent on the spatial resolution of the predator–prey models in use. To demonstrate the sensitivity of the model results on economic parameters and spatial resolution, the standard analytical description of population dynamics and economic adjustments is combined with a dynamic modeling approach. Although economic models typically concentrate on the analytical investigation of properties of a system's equilibrium points, the dynamic modeling approach chosen for this chapter admits the richness of experimental investigations to models of the dynamic evolution of fishery resources. This chapter demonstrates again the power of merging experimentation with economic models of natural resource use to assess the sensitivity of model results to model specifications, initial conditions, and empirical estimates and to guide data collection, analysis, and ultimately policy decision making.

Assume the following dynamics of the stock N of a fish population in a given area:

$$\frac{dN}{dt} = \dot{N} = R^*N^*\left(1 - \frac{N}{K}\right) - V^*E^*N = R^*N^*\left(1 - \frac{N}{K}\right) - Q, \qquad (27.1)$$

with R as the natural rate of increase of the population; K, the carrying capacity of the ecosystem, measured in tons of biomass; E, effort, measured in days fished in the fishery in a given year; and V, a catchability coefficient characteristic of the technology in use and measured in tons of biomass caught per biomass in the fishery per fishing day per time period. The product $Q = V^*E^*N$ is total catch in the fishery per time period, and thus, $P^*V^*E^*N$ is revenue from the catch. For simplicity, the carrying capacity, catachability coefficients, and price are assumed constant.

Assume for the fishermen that all profit made in a given period is used to increase effort. Thus,

$$\frac{dE}{dt} = \dot{E} = E^*(V^*P^*N - J), \qquad (27.2)$$

with P as price per ton of biomass and J as cost per unit effort, including costs of operating boats, paying fishermen, and reimbursing the captain. Cost per unit

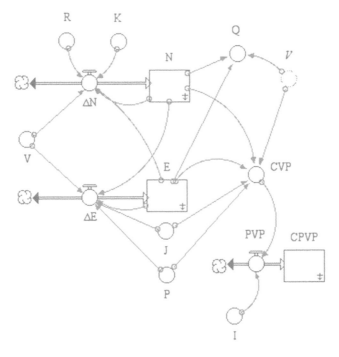

Fig. 27.1 Basic preditor–prey model of a fishery

effort are measured in dollars per day fished. For simplicity, we assume the price to be constant.

The goal of the fishermen is to maximize the cumulative present value of profits:

$$\text{CPCP} = \int_0^T E^*(V^*P^*N - J)\, e^{-I^*t}\mathrm{d}t. \qquad (27.3)$$

The interest rate, I, is assumed to be constant.

The optimal harvest Q of the fish population is determined by the optimal effort level. That level can be found by choosing an initial effort level, running the model, noting the resulting cumulative present value of profits, and adjusting effort in consecutive runs until an initial effort level is found that yields the highest possible cumulative present value of profits. The model is shown in Fig. 27.1.

Assume the following parameters: The initial population and carrying capacity are 40,000 and 150,000 tons of biomass, respectively. The natural rate of growth of the fish population, in the absence of fishing, is 0.08 tons of biomass per ton of biomass per time period. Costs per unit effort are \$0.22 per fishing day per ton, price per ton of fish is \$2.9. The catchability coefficient is 0.000008 tons of biomass per ton of fish in the fishery per time period. At an interest rate of 5%, in the initial year the optimal level of effort is 992 fishing days.

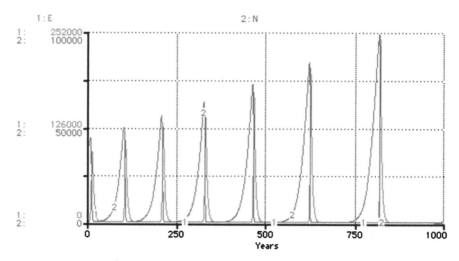

Fig. 27.2 Time series for effort and population size

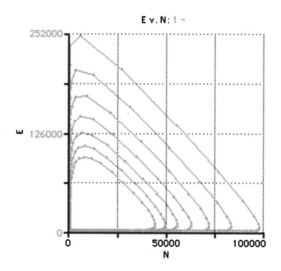

Fig. 27.3 Observed changes in effort with changes in population size

This model, when run over centuries (DT = 0.5), yields oscillations that tend to increase beyond bounds. Each increase in the fish population prompts the fishermen to increase their effort and decreases in fish population result in decreased effort due to reduced profits (Figs. 27.2 and 27.3). Note also that the present value of profits can temporarily be negative. Thus, to find the optimal initial effort level, one must run the model for a sufficiently long time period; that is, until present value profits become discounted so much that their contribution to CPVP is negligible (Fig. 27.4).

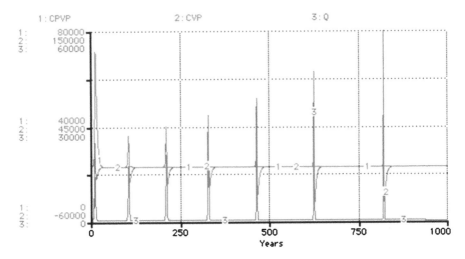

Fig. 27.4 Profits and catch

Extend the model to account for price changes along a linear demand curve. Choose a demand curve that has the price $P = 2.9$ at a catch $Q = 318.4$ tons of biomass as one of its points, so that you can start your model at the same point as in the previous one. Be careful not to have price changes too sensitive to changes in catch. How do optimal initial effort and cumulative present value of profits change when such a demand curve is introduced? How does a harvest tax affect optimal effort? Be sure to speculate on the form of the answer before you run the model.

27.2   Basic Fisheries Model Equations

CPVP(t) = CPVP(t - dt) + (PVP) * dt
INIT CPVP = 0
INFLOWS:
PVP = CVP*EXP(-I*TIME)
E(t) = E(t - dt) + (ΔE) * dt
INIT E = 992 {Fishing days}
INFLOWS:
ΔE = E*(V*P*N-J) {fishing days per time period}
N(t) = N(t - dt) + (ΔN) * dt
INIT N = 40000 {Tons of fish}
INFLOWS:
ΔN = R*N*(1-N/K) - V*E*N {Tons of fish per ton of fish in fishery per time period}
CVP = E*(V*P*N-J)
I = .05 {Interest rate; $ per $ per year}
J = .22 {$ per fishing day per ton of fish}
K = 150000 {carrying capacity; tons of fish in fishery}
P = 2.9 {$ per ton of fish}
Q = E*V*N
R = .08 {new tons of fish per period per ton of fish in fishery}
V = .000008 {tons of fish caught per tons of fish in fishery per time period}

27.3   Fishing with Nonmalleable Capital

In reality, it is not as easy for fishermen to adjust effort as we assumed in the previous section. Capital equipment, for example, needs to be bought or sold. Thus, we may want to include the cost of new capital expenditures and revenues that arise from selling old capital. Let us for simplicity assume that the price of new and old capital is the same. As in the model of the previous section $P*V*E*N$ are the revenues from fishing (dollars per time period). Denote B as the stock of capital at a period of time, such as the number of boats owned and operated by the fishermen. $\dot{B}$ is then the change in boats per time period. To calculate costs, we need to distinguish those that are directly related to operating the capital equipment (boats, mainly) and those of purchasing new boats. Total operating costs are $J*B$ (dollars per fishing day). With P_B as the price of capital, $P_B*\dot{B}$ is the cost of purchasing new capital (i.e., if $\dot{B}>0$). Similarly, if $\dot{B}<0$ then old boats are sold and since we assume here for simplicity that the price of old boats and new ones is the same, $P_B*\dot{B}$ is revenues from selling old capital. Current value of profits, CVP, is then

$$\text{CVP} = E^*V^*P^*N - J^*B - P_B{}^*\dot{B}. \tag{27.4}$$

Let us also assume that effort is proportional to the amount of capital owned by the fishermen:

$$E = A^*B, \tag{27.5}$$

$$\dot{E} = A^*\dot{B}. \tag{27.6}$$

The proportionality factor A is measured in fishing days per unit of capital. For all intents and purposes we may normalize $A = 1$ and replace $\dot{B}$ by $\dot{E}$ in CVP:

$$\text{CVP} = E^*V^*P^*N - J^*E - P_B{}^*\dot{E}. \tag{27.7}$$

Hence, the change in effort in a given year, $\dot{E}$, is the following function in profits

$$\dot{E} = E^*(V^*P^*N - J) - P_B{}^*\dot{E} \tag{27.8}$$

or

$$\dot{E} = \frac{1}{1 + P_B} E^*(V^*P^*N - J) \tag{27.9}$$

or

$$\dot{E} = \text{ALPHA}^*E(V^*P^*N - J) \tag{27.10}$$

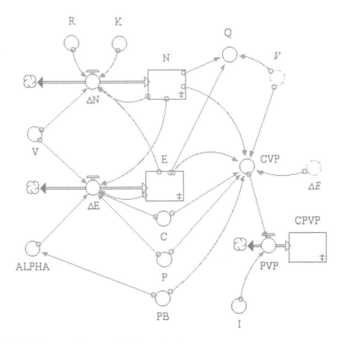

Fig. 27.5 Basic fishery with nonmalleable capital

with

$$\text{ALPHA} = \frac{1}{1 + P_B}.\tag{27.11}$$

For $P_B = 0$ (ALPHA $= 1$) we have the case in which capital can be acquired at no cost and is given away without generating revenues. This is the case modeled in the previous section. Capital is perfectly malleable; that is, it can be used and retired at any time and without any constraints.

Set $P_B = 3$ in the model of Fig. 27.5. It is now more difficult for the fishermen to adjust their effort (ALPHA $= 0.25$) because they need to spend a portion of their profits on new capital. In return, however, they generate revenues when they sell capital.

Run the model at DT $= 0.5$. In contrast to the model of the previous section with ALPHA $= 1$, the resulting levels of effort and the fish population sizes show damped oscillations (Figs. 27.6–27.8). The optimal initial effort level is 989 fishing days, resulting in a cumulative present value of profits of approximately $14,799.87.

In the previous section we found that, for ALPHA $= 1$, the system exhibits explosive oscillations; that is, over time both effort and population levels increase beyond bounds. For ALPHA $= 0.25$, the system in the long run will move toward a steady state. Hence, there must be a value for ALPHA at which the system will

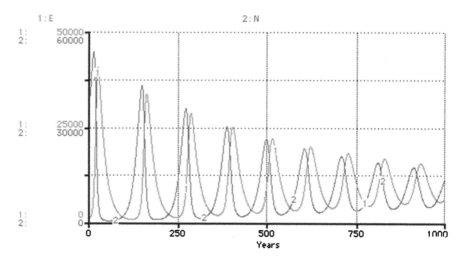

Fig. 27.6 Time series for effort and population size with nonmalleable capital

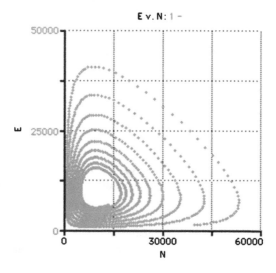

Fig. 27.7 Observed changes in effort with changes in population size (ALPHA = 0.25)

repeat its behavior over and over again, a so-called stable limit cycle. Lower adjustment coefficients ALPHA tend to damp oscillations because density dependence of the fish population dynamics increasingly dominates the system's behavior.

Slightly increase ALPHA for consecutive runs. You should find, for an initial effort of 990 fishing days, values of ALPHA that result in explosive oscillations, damped oscillations, or neutral stability. Is there a unique combination of values for ALPHA

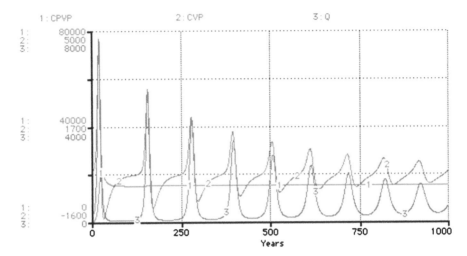

Fig. 27.8 Profits and catch

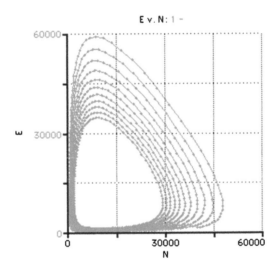

Fig. 27.9 Observed changes in effort with changes in population size (ALPHA = 0.50)

and initial effort that yields a stable limit cycles? The cycles for $E(t = 0) = 990$ and ALPHA = 0.50, ALPHA = 0.66, ALPHA = 0.75 are shown, respectively, in Figs. 27.9–27.11.

Of course, real adjustment coefficients ALPHA depend on the prices of capital (e.g., gear, boats) of different vintages, interest rates, institutional constraints (e.g., tax breaks, subsidies, limitations of access to the fishery, restrictions on boat size, or fishing seasons) and characteristics of fishermen (e.g., risk aversion, price expectations, experience). Thus, changes in ALPHA are very likely and potentially

Fig. 27.10 Observed
changes in effort with
changes in population size
(ALPHA = 0.66)

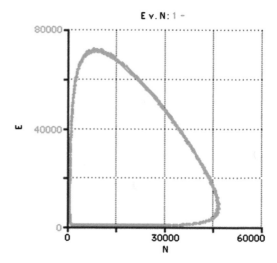

Fig. 27.11 Observed
changes in effort with
changes in population size
(ALPHA = 0.75)

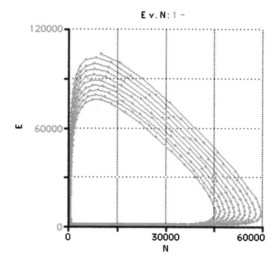

perturbing the system's dynamics. Even small changes in ALPHA, as it is evident from the models developed in this chapter, result in marked differences of the system behavior. Little data is actually available to reliably estimate ALPHA. The resulting models are consequently highly sensitive to the assumptions about the actual adjustments in effort, and it is therefore difficult to attribute population dynamics predominantly to environmental fluctuations.

Return to the model with ALPHA = 0.25 and generalize the model to have purchase prices of new capital higher than the sales prices of "scrap," *PS*. Introduce a depreciation rate for the capital stock. How do a harvest tax and a property tax (a tax per boat) affect optimal effort?

27.4 Nonmalleable Capital Stock Model Equations

CPVP(t) = CPVP(t - dt) + (PVP) * dt
INIT CPVP = 0
INFLOWS:
PVP = CVP*EXP(-I*TIME)
E(t) = E(t - dt) + (ΔE) * dt
INIT E = 989 {Fishing days}
INFLOWS:
ΔE = ALPHA*E*(V*P*N-C) {fishing days per time period}
N(t) = N(t - dt) + (ΔN) * dt
INIT N = 40000 {Tons of fish}
INFLOWS:
ΔN = R*N*(1-N/K) - V*E*N {Tons of fish per ton of fish in fishery per time period}
ALPHA = 1/(1+PB)
C = .22 {$ per fishing day per ton of fish}
CVP = E*(V*P*N-C) - PB*ΔE
I = .05 {Interest rate; $ per $ per year}
K = 150000 {carrying capacity; tons of fish in fishery}
P = 2.9 {$ per ton of fish}
PB = 3 {$ per fishing day}
Q = E*V*N
R = .08 {new tons of fish per period per ton of fish in fishery}
V = .000008 {tons of fish caught per tons of fish in fishery per time period}

References

Ruth M (1995) A system dynamics approach to modeling fisheries management issues: applications for spatial dynamics and resolution. Syst Dyn Rev 11:233–243

Swart J (1990) A system dynamics approach to predator–prey modeling. Syst Dyn Rev 6:94–99

Wilson JA, Townsend R, Kelban P, McKay S, French J (1990) Managing unpredictable resources: traditional policies applied to chaotic populations. Ocean Shoreline Manag 13:179–197

Chapter 28
Spatial Fishery Model

> *Then I say the earth belongs to each ... generation during its*
> *course, fully and in its own right. ... Then, no generation can*
> *contract debts greater than may be paid during the course of*
> *its own existence.*

Thomas Jefferson, September 6, 1789

28.1 Basic Model

In the previous chapter, we saw that, to a significant degree, fish population dynamics may be attributed to economic factors. This is not surprising, as humans are frequently the number one predator of fish species. Consequently, fisheries models must incorporate very carefully economic adjustment processes. In this chapter, we show that differences between the model result and actual system behavior are also significantly dependent on the spatial resolution of the predator–prey models in use. Allen and McGlade (1987) make use of a more elaborate spatial fisheries model, including a specification of effort adjustments that is similar to the one employed here. Their model, however, is devoted to the macroscopic results of adjustments in individual fishermen's risk perception, and does not address the issues of spatial resolution discussed here.

Let us investigate the implications of spatial resolution for the model results. Toward this end, we modify the general predator–prey model of the previous chapter for a marine fishery subdivided into n regions. The change in population in region i ($i = 1{:}n$) is given by

A save-disabled version of STELLA® and the computer models of this book are available at www.iseesystems.com/modelingeconomicsystems.

M. Ruth and B. Hannon, *Modeling Dynamic Economic Systems*,
Modeling Dynamic Systems, DOI 10.1007/978-1-4614-2209-9_28,
© Springer Science+Business Media, LLC 2012

$$\frac{\mathrm{d}N}{\mathrm{d}t} = \dot{N_t} = R * N_i * \left(1 - \frac{N_i}{K_i}\right) - V * E_i * N_i, \tag{28.1}$$

with R as the natural rate of change in the absence of fishing-induced mortality; K_i, the carrying capacity of region i; V, a constant catchability coefficient (dependent on technology); and E_i, fishing effort in region i. Hence, $V*E_i*N_i$ is the total harvest from region i:

$$Q_i = V * E_i * N_i \tag{28.2}$$

and

$$Q = \sum_{i=1}^{n} Q_i. \tag{28.3}$$

For simplicity we may assume all regions to be of equal size and quality; that is,

$$K_i = K_j = \text{constant.} \tag{28.4}$$

Changes in total effort E are given by

$$\frac{\mathrm{d}E}{\mathrm{d}t} = \dot{E} = \alpha * (P * V * E * N - J * E), \tag{28.5}$$

with α a constant proportionality factor, P the unit price of a catch, N total population, and J the unit cost per effort. Total effort is then allocated to region i based on the relative population size D_i of that region; that is, $E_i = D_i*E$. Typically, $\alpha < 1$ due to limited malleability of capital in the fishing industry.

The STELLA model (Fig. 28.1) is composed of several modules. In one such module, we calculate the stock of fish biomass.

A second module (Fig. 28.2) is set up to allocate effort to each of the four regions. In this module, we calculate the proportions D_i of fish biomass in region i ($i = 1, 2, 3, 4$) and then multiply total effort by those proportions. In the model without migration (setting $M = 0$), equal distribution of biomass across regions, and equal size and quality of the regions, D_i will always be 0.25. Therefore, 25% of the effort is directed to each of the four regions. If we start out the model with unequal distributions of fish or with unequal size ($K_i \neq K_j$), the D_i typically will not be the same.

Total harvest in each region and cumulative present value of profits are calculated in a third module (Fig. 28.3). It is set up to account for differences in costs per unit of effort for different regions. Once you run the basic model, you may want to investigate the impact of costs differing by region, say, due to different restrictions on gear or boats in different fishing grounds. You may then also want to change the decision on effort allocation among regions, shown in the previous module.

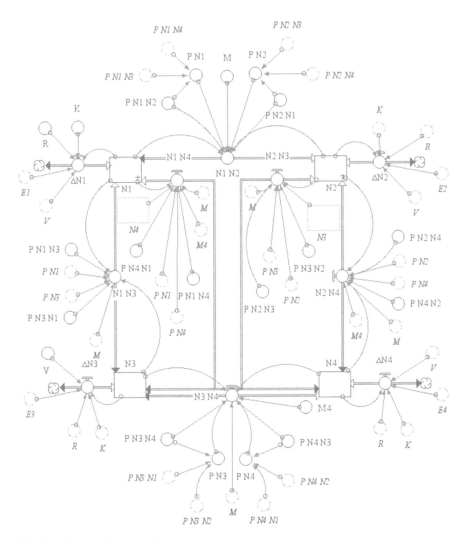

Fig. 28.1 Spatial fishery model

Let us start with 5,000 tons of biomass fish in each region and a carrying capacity of 20,000 tons of biomass in each region. To run the model, let us assume that the size of the fishing fleet at first is small, enabling the fish population to increase initially. This increase will then prompt an increase in effort. If we assume identical initial conditions for each region and no migration of fish across regional boundaries, the specification is equivalent to a single region model, as we discussed it in the previous chapter.

To verify the existence of damped and explosive oscillations and neutral stability in the absence of migration, set $M = 0$ and E $(t = 0) = 150$ and run the model

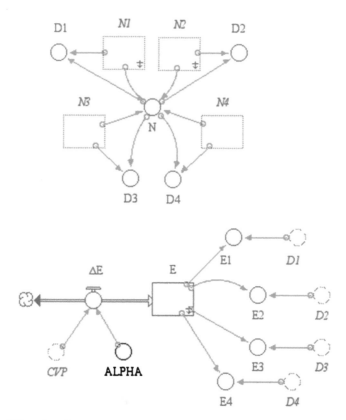

Fig. 28.2 Effort allocation across regions

with $\alpha = 0.60$, $\alpha = 0.6666$, and $\alpha = 0.70$ for a sufficiently long time (e.g., 1,000 periods). Run the model at DT $= 0.5$. What you should get for $\alpha = 0.6666$ is illustrated in Figs. 28.4–28.7.

To see the implications of changes in the spatial resolution for the model results, include the spatial aspects of the four-region fishery by allowing for migration of fish among regions. Assume that a fixed proportion of the population migrates, and the destination of those fish is determined by random factors. No fish, however, are leaving the system other than through natural or fishing-induced death. None are added other than through births.

Assume again an equal initial distribution of fish populations across the four regions. The regions are contiguously arranged in a square and migration of fish is permitted from each region into any other. As a result, the relative size of the fish population in each area changes from time step to time step. Thus, typically, $D_i \neq D_j$ and therefore effort is not uniform across the regions.

Net migration N_iN_j between regions i and j ($i \neq j$) is calculated as

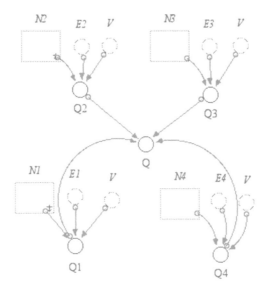

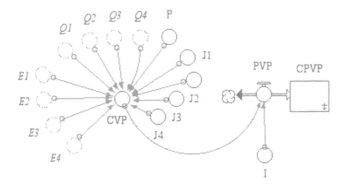

Fig. 28.3 Harvest and profits

$$N_i N_j = M_i N_i \frac{P_{N_i N_j}}{\sum_i P_{N_i N_j}} - M_j N_j \frac{P_{N_j N_i}}{\sum_j P_{N_j N_i}}, \tag{28.6}$$

with M_i and M_j as the proportion of migrants from regions i and j, respectively; and $P_{N_i N_j}$ and $P_{N_j N_i}$, random numbers ranging between 0 and 1. $N_1 N_2$, for example, is the net migration of fish (measured in tons of fish biomass) between regions 1 and 2. Set all initial conditions and parameters equal to the ones that yielded neutral stability in the model runs of the previous chapter.

Assuming that fish move freely in oceans is not at all an unreasonable assumption. To the observer (or modeler) the proportion of fish moving from one region to

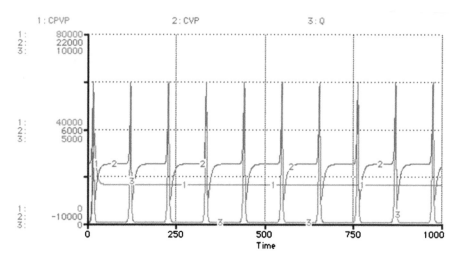

Fig. 28.4 Profits and catch

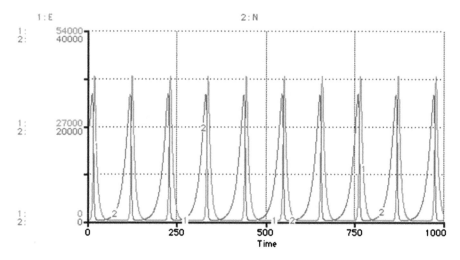

Fig. 28.5 Effort and population size

another may seem higher the finer the spatial resolution. Increase of the spatial resolution from the one-region to a four-region model results in movement away from neutral stability to explosive oscillations. The terms *neutral stability* and *oscillation* are used here loosely because the randomness of migration will always introduce fluctuations in the time paths.

With increased migration rates M_i, the oscillations become more pronounced. Figures 28.8–28.11 assume, for simplicity, $M_i = M_j = M = 0.90, \forall i, j$. You can

Fig. 28.6 Effort levels
at different population sizes

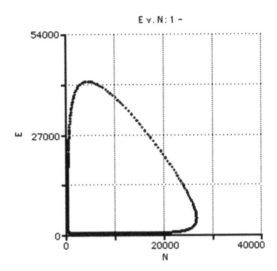

Fig. 28.7 Changes
in population for different
population sizes

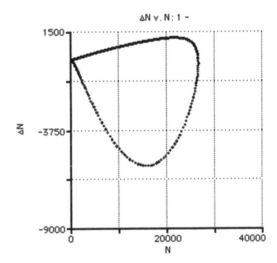

easily relax this assumption. Additionally, we assume $\alpha = 0.6666$; that is, the adjustment coefficient yields neutral stability in the absence of migration.

By increasing the spatial resolution of the model and enabling movement among the regions of the model, and leaving virtually all other model parameters unchanged, we no longer generate neutral stability but explosive oscillations. Additionally, CPVP is lower than in the single-region model and lower for higher values of M_i. Is this conclusion robust for alternative arrangements of the four regions and for larger numbers of regions?

Organize the model so that the regions are in a row rather than a square. Alternatively, prevent diagonal migration in the four-region model. Increase the

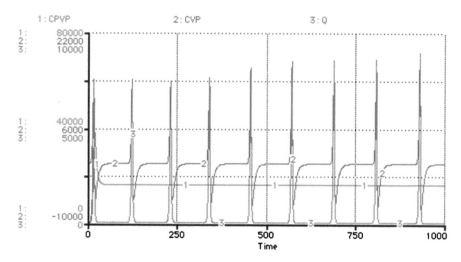

Fig. 28.8 Profits and catch with migration

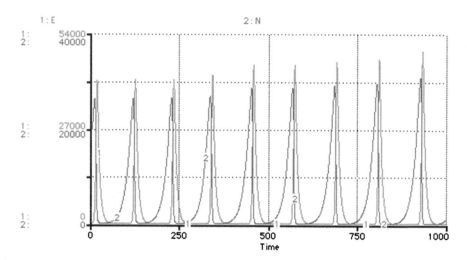

Fig. 28.9 Effort and population size with migration

number of regions to 9, then to 16, and simulate the models with the parameters that yield a stable limit cycle in the absence of migration. What are the implications of an unequal initial distribution of biomass across regions or differences in the size (carrying capacity) of the regions for the system's dynamics. Can you provide a general statement for the model behavior as the spatial resolution increases? What are the implications for the optimal initial effort level?

Fig. 28.10 Effort levels
at different population sizes
with migration

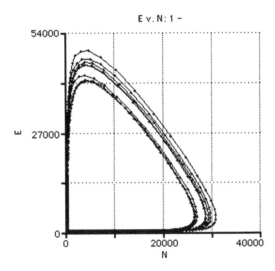

Fig. 28.11 Changes
in population for different
population sizes with
migration

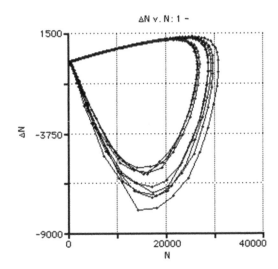

Let the boats follow the fish as they flourish in neighboring regions, with a lag time. Introduce technical change that, for example, leads to an increase in the catchability coefficient over time. Introduce a demand curve. Carry out each of these modifications individually and, before running the model, make an educated guess about the results.

28.2   Spatial Fisheries Model Equations

CPVP(t) = CPVP(t - dt) + (PVP) * dt
INIT CPVP = 0
INFLOWS:
PVP = CVP*EXP(-I*TIME)
E(t) = E(t - dt) + (ΔE) * dt
INIT E = 150 {Fishing hours per tons of fish}
INFLOWS:
ΔE = ALPHA*CVP
N1(t) = N1(t - dt) + (ΔN1 + N1_N3 + N1_N4 - N1_N2) * dt
INIT N1 = 5000 {Tons of fish}
INFLOWS:
ΔN1 = R*N1*(1-N1/K) - V*E1*N1 {Tons of fish per ton of fish in
fishery per time period}
N1_N3 = (N3*M*P_N3_N1/P_N3) - (N1*M*P_N1_N3/P_N1)
{Netmigration; tons per year}
N1_N4 = (N4*M4*P_N4_N1/P_N4) - (N1*M*P_N1_N4/P_N1)
{Netmigration; tons per year}
OUTFLOWS:
N1_N2 = (N1*M*P_N1_N2/P_N1) -(N2*M*P_N2_N1/P_N2)
{Netmigration; tons per year}
N2(t) = N2(t - dt) + (ΔN2 + N1_N2 + N2_N4 + N2_N3) * dt
INIT N2 = 5000 {Tons of fish}
INFLOWS:
ΔN2 = R*N2*(1-N2/K) - V*E2*N2 {Tons of fish per ton of fish in
fishery per time period}
N1_N2 = (N1*M*P_N1_N2/P_N1) -(N2*M*P_N2_N1/P_N2)
{Netmigration; tons per year}
N2_N4 = (N4*M4*P_N4_N2/P_N4) - (N2*M*P_N2_N4/P_N2)
{Netmigration; tons per year}
N2_N3 = (N3*M*P_N3_N2/P_N3) - (N2*M*P_N2_N3/P_N2)
{Netmigration; tons per year}
N3(t) = N3(t - dt) + (ΔN3 - N3_N4 - N1_N3 - N2_N3) * dt
INIT N3 = 5000 {Total biomass; tons}
INFLOWS:
ΔN3 = R*N3*(1-N3/K) - V*E3*N3 {Tons of fish per ton of fish in
fishery per time period}
OUTFLOWS:
N3_N4 = (N3*M*P_N3_N4/P_N3) -(N4*M4*P_N4_N3/P_N4)
{Netmigration; tons per year}
N1_N3 = (N3*M*P_N3_N1/P_N3) - (N1*M*P_N1_N3/P_N1)
{Netmigration; tons per year}
N2_N3 = (N3*M*P_N3_N2/P_N3) - (N2*M*P_N2_N3/P_N2)
{Netmigration; tons per year}
N4(t) = N4(t - dt) + (ΔN4 + N3_N4 - N2_N4 - N1_N4) * dt
INIT N4 = 5000 {Total biomass; tons}
INFLOWS:
ΔN4 = R*N4*(1-N4/K) - V*E4*N4 {Tons of fish per ton of fish in
fishery per time period}
N3_N4 = (N3*M*P_N3_N4/P_N3) -(N4*M4*P_N4_N3/P_N4)
{Netmigration; tons per year}
OUTFLOWS:

N2_N4 = (N4*M4*P_N4_N2/P_N4) - (N2*M*P_N2_N4/P_N2)
{Netmigration; tons per year}
N1_N4 = (N4*M4*P_N4_N1/P_N4) - (N1*M*P_N1_N4/P_N1)
{Netmigration; tons per year}
ALPHA = .6666 {fishing hours per ton of fish}
CVP = P*(Q1+Q2+Q3+Q4)-J1*E1-J2*E2-J3*E3-J4*E4
D1 = N1/(N) {Relative density in region 1; tons per ton}
D2 = N2/(N) {Relative density in region 2; tons per ton}
D3 = N3/(N) {Relative density in region 3; tons per ton}
D4 = N4/N {Relative density in region 4; tons per ton}
E1 = D1*E {Fishing hours per tons of fish}
E2 = D2*E {Fishing hours per tons of fish}
E3 = D3*E {Fishing hours per tons of fish}
E4 = D4*E {Fishing hours per tons of fish}
I = .05
J1 = .22 {$ per fishing hour per ton of fish}
J2 = .22 {$ per fishing hour per ton of fish}
J3 = .22 {$ per fishing hour per ton of fish}
J4 = .22 {$ per fishing hour per ton of fish}
K = 20000 {carrying capacity; tons of fish in fishery}
M = 0 {Migration rate; tons per ton per time period}
M4 = 0 {Migration rate; tons per ton per year}
N = N1+N2+N3+N4 {Total biomass; tons}
P = 2.9 {$ per ton of fish}
P_N1 = P_N1_N2 + P_N1_N3 +P_N1_N4
P_N1_N2 = RANDOM(0,1) {Probability of moving from N1 to N2}
P_N1_N3 = RANDOM(0,1) {Probability of moving from N1 to N3}
P_N1_N4 = RANDOM(0,1) {Probability of moving from N1 to N4}
P_N2 = P_N2_N1 + P_N2_N3 + P_N2_N4
P_N2_N1 = RANDOM(0,1) {Probability of moving from N2 to N1}
P_N2_N3 = RANDOM(0,1) {Probability of moving from N1 to N2}
P_N2_N4 = RANDOM(0,1) {Probability of moving from N2 to N4}
P_N3 = P_N3_N4 + P_N3_N2 + P_N3_N1
P_N3_N1 = RANDOM(0,1) {Probability of moving from N3 to N1}
P_N3_N2 = RANDOM(0,1) {Probability of moving from N1 to N4}
P_N3_N4 = RANDOM(0,1) {Probability of moving from N1 to N2}
P_N4 = P_N4_N3 + P_N4_N2 + P_N4_N1
P_N4_N1 = RANDOM(0,1) {Probability of moving from N1 to N4}
P_N4_N2 = RANDOM(0,1) {Probability of moving from N4 to N2}
P_N4_N3 = RANDOM(0,1) {Probability of moving from N2 to N1}
Q = Q1+Q2+Q3+Q4 {Total harvest; tons per year}
Q1 = V*E1*N1 {Harvest; tons per year}
Q2 = V*E2*N2 {Harvest; tons per year}
Q3 = V*E3*N3 {Harvest; tons per year}
Q4 = V*E4*N4 {Harvest; tons per year}
R = .08 {new tons of fish per period per ton of fish in fishery}
V = .00006 {tons of fish caught per tons of fish in fishery per time
period}
ΔN = ΔN1+ΔN2+ΔN3+ΔN4

28.3   Management of a Multiregion Fishery

Previous fisheries policies targeted individual species through various restrictions on effort or catch. In lieu of sufficient understanding of the feedbacks among fish populations and their physical environment as well as the corresponding economic adjustments, management may more safely be done by setting aside reserves (Holland 1993). No fishing would be allowed in the reserve but fish could leave the reserve, thus replenishing stocks in the remainder of the ocean. In this section, we introduce a reserve into the spatial fisheries model of the previous section.

The STELLA model for the four-region model with one region designated as a reserve is virtually the same as in the previous section. Just modify the module that allocates effort to the regions so that no fishing takes place in the reserve (e.g., set $E_4 = 0$ if region 4 is the reserve) and distribute effort such that all effort is allocated to the other three regions.

To better understand the effects of the fishery reserve, let us first assume that no migration takes place across any of the regional boundaries. The result is, of course, a higher total population level than in the absence of the reserve. Once the stable oscillation is reached, population never drops below the size maintained by the reserve. As a result, the system moves from its initial oscillation (left part of the following two graphs) to the stable limit cycle. Although the population size increases, the cumulative present value of profits is lower because part of the fish population is off limits to fishermen (Figs. 28.12 and 28.13).

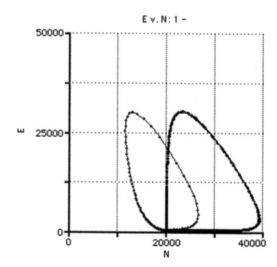

Fig. 28.12 Effort levels at different population sizes with a reserve and no migration

Fig. 28.13 Changes in
population for different
population sizes with a
reserve and no migration

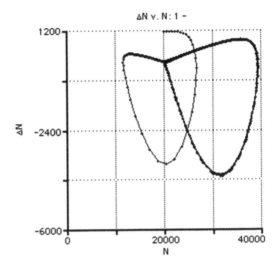

Fig. 28.14 Effort levels
at different population sizes
with a reserve and differing
migration rates

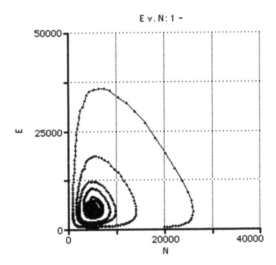

Now let us include migration in the model. Assume that the reserve is placed to cover the region with highest rates of migration; that is, $M_4 > M_1 = M_2 = M_3$. For example, assume that 90% of the fish in the reserve (region 4) migrate and only 50% in the remaining region. Again, start the model with $\alpha = 0.6666$, an effort level of 150 fishing days and an initial biomass of 5,000 tons in each of the four regions. The carrying capacity of each of these regions is $K_i = K = 20,000$.

Fig. 28.15 Changes
in population for different
population sizes with
a reserve and differing
migration rates

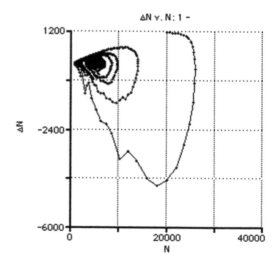

As Figs. 28.14 and 28.15 attest, migration leads to abandonment of the stable limit cycle but, unlike in the absence of the reserve, to damped oscillations. Due to migration, the population size may fall below the maximum that can be sustained in the reserve. To judge about the change in cumulative present value of profits when migration is introduced in the model, you need to run the model many times, because the randomness of the migration will result in different CPVP values for consecutive runs. Typically, the cumulative present value of profits is marginally lower than in the one region model with stable limit cycle.

Set up the four-region, one-reserve model such that the regions are not arranged in a square but a row; for example, we may be modeling a narrow fjord rather than a wide bay. You should find a faster movement toward the steady state, and cumulative present values of profit that may exceed those achieved in the one-region model with neutral stability. The rather erratic initial development is caused by the combined effect of the randomness of the migration direction of fish from the reserve and its immediate neighbor region(s).

The spatial model shows that, starting from neutral stability, an increase in spatial resolution and an introduction of the reserve leads to fundamentally different results with regard to long-term population sizes and economic performance. Therefore, when modeling the optimal use of a mobile renewable resource, special attention must be paid not only to the biological characteristics of the resource but also to the spatial resolution at which the system is modeled. To provide qualitative, let alone quantitative, answers on the dynamics of fish populations and effort requires clarification of the appropriate spatial resolution of models and proper delineation of the regions to be modeled.

28.4   Multiregion Fisheries Model Equations

CPVP(t) = CPVP(t - dt) + (PVP) * dt
INIT CPVP = 0
INFLOWS:
PVP = CVP*EXP(-I*TIME)
E(t) = E(t - dt) + (ΔE) * dt
INIT E = 150 {Fishing hours per tons of fish}
INFLOWS:
ΔE = ALPHA*CVP
N1(t) = N1(t - dt) + (_N1 + N1_N3 + N1_N4 - N1_N2) * dt
INIT N1 = 5000 {Tons of fish}
INFLOWS:
ΔN1 = R*N1*(1-N1/K) - V*E1*N1 {Tons of fish per ton of fish in
fishery per time period}
N1_N3 = 0*((N3*M*P_N3_N1/P_N3) -
(N1*M*P_N1_N3/P_N1)) {Netmigration; tons per year}
N1_N4 = 0*((N4*M4*P_N4_N1/P_N4) -
(N1*M*P_N1_N4/P_N1)) {Netmigration; tons per year}
OUTFLOWS:
N1_N2 = (N1*M*P_N1_N2/P_N1) -(N2*M*P_N2_N1/P_N2)
{Netmigration; tons per year}
N2(t) = N2(t - dt) + (_N2 + N1_N2 + N2_N4 + N2_N3) * dt
INIT N2 = 5000 {Tons of fish}
INFLOWS:
ΔN2 = R*N2*(1-N2/K) - V*E2*N2 {Tons of fish per ton of fish in
fishery per time period}
N1_N2 = (N1*M*P_N1_N2/P_N1) -(N2*M*P_N2_N1/P_N2)
{Netmigration; tons per year}
N2_N4 = 0*((N4*M4*P_N4_N2/P_N4) -
(N2*M*P_N2_N4/P_N2)) {Netmigration; tons per year}
N2_N3 = (N3*M*P_N3_N2/P_N3) - (N2*M*P_N2_N3/P_N2)
{Netmigration; tons per year}
N3(t) = N3(t - dt) + (_N3 - N3_N4 - N1_N3 - N2_N3) * dt
INIT N3 = 5000 {Total biomass; tons}
INFLOWS:
ΔN3 = R*N3*(1-N3/K) - V*E3*N3 {Tons of fish per ton of fish in
fishery per time period}
OUTFLOWS:
N3_N4 = (N3*M*P_N3_N4/P_N3) -(N4*M4*P_N4_N3/P_N4)
{Netmigration; tons per year}
N1_N3 = 0*((N3*M*P_N3_N1/P_N3) -
(N1*M*P_N1_N3/P_N1)) {Netmigration; tons per year}
N2_N3 = (N3*M*P_N3_N2/P_N3) - (N2*M*P_N2_N3/P_N2)
{Netmigration; tons per year}
N4(t) = N4(t - dt) + (_N4 + N3_N4 - N2_N4 - N1_N4) * dt
INIT N4 = 5000 {Total biomass; tons}
INFLOWS:
ΔN4 = R*N4*(1-N4/K) - V*E4*N4 {Tons of fish per ton of fish in
fishery per time period}
N3_N4 = (N3*M*P_N3_N4/P_N3) -(N4*M4*P_N4_N3/P_N4)
 {Netmigration; tons per year}

OUTFLOWS:

N2_N4 = 0*((N4*M4*P_N4_N2/P_N4) -
(N2*M*P_N2_N4/P_N2)) {Netmigration; tons per year}

N1_N4 = 0*((N4*M4*P_N4_N1/P_N4) -
(N1*M*P_N1_N4/P_N1)) {Netmigration; tons per year}

ALPHA = .6666 {fishing hours per ton of fish}

CVP = P*(Q1+Q2+Q3+Q4)-J1*E1-J2*E2-J3*E3-J4*E4

D1 = N1/(N-N4) {Relative density in region 1; tons per ton}

D2 = N2/(N-N4) {Relative density in region 2; tons per ton}

D3 = N3/(N-N4) {Relative density in region 3; tons per ton}

D4 = 0*N4/N {Relative density in region 4; tons per ton}

E1 = D1*E {Fishing hours per tons of fish}

E2 = D2*E {Fishing hours per tons of fish}

E3 = D3*E {Fishing hours per tons of fish}

E4 = 0*D4*E {Fishing hours per tons of fish}

I = .05

J1 = .22 {$ per fishing hour per ton of fish}

J2 = .22 {$ per fishing hour per ton of fish}

J3 = .22 {$ per fishing hour per ton of fish}

J4 = .22 {$ per fishing hour per ton of fish}

K = 20000 {carrying capacity; tons of fish in fishery}

M = .5 {Migration rate; tons per ton per time period}

M4 = .9 {Migration rate; tons per ton per year}

N = N1+N2+N3+N4 {Total biomass; tons}

P = 2.9 {$ per ton of fish}

P_N1 = P_N1_N2 + P_N1_N3 +P_N1_N4

P_N1_N2 = RANDOM(0,1) {Probability of moving from N1 to N2}

P_N1_N3 = RANDOM(0,1) {Probability of moving from N1 to N3}

P_N1_N4 = RANDOM(0,1) {Probability of moving from N1 to N4}

P_N2 = P_N2_N1 + P_N2_N3 + P_N2_N4

P_N2_N1 = RANDOM(0,1) {Probability of moving from N2 to N1}

P_N2_N3 = RANDOM(0,1) {Probability of moving from N1 to N2}

P_N2_N4 = RANDOM(0,1) {Probability of moving from N2 to N4}

P_N3 = P_N3_N4 + P_N3_N2 + P_N3_N1

P_N3_N1 = RANDOM(0,1) {Probability of moving from N3 to N1}

P_N3_N2 = RANDOM(0,1) {Probability of moving from N1 to N4}

P_N3_N4 = RANDOM(0,1) {Probability of moving from N1 to N2}

P_N4 = P_N4_N3 + P_N4_N2 + P_N4_N1

P_N4_N1 = RANDOM(0,1) {Probability of moving from N1 to N4}

P_N4_N2 = RANDOM(0,1) {Probability of moving from N4 to N2}

P_N4_N3 = RANDOM(0,1) {Probability of moving from N2 to N1}

Q = Q1+Q2+Q3+Q4 {Total harvest; tons per year}

Q1 = V*E1*N1 {Harvest; tons per year}

Q2 = V*E2*N2 {Harvest; tons per year}

Q3 = V*E3*N3 {Harvest; tons per year}

Q4 = V*E4*N4 {Harvest; tons per year}

R = .08 {new tons of fish per period per ton of fish in fishery}

V = .00006 {tons of fish caught per tons of fish in fishery per
time period}

ΔN = ΔN1+ΔN2+ΔN3+ΔN4

References

Allen P, McGlade JM (1987) Modelling complex human systems: a fisheries example. Eur J Oper Res 30:147–167

Holland D (1993) Managing fisheries without restricting catch or effort: the use of marine reserves for inshore fisheries. MS Thesis, University of Illinois, Department of Agricultural Economics

Part VI
Chaos in Economic Models

Chapter 29
Preference Cycles and Chaos

It could be argued that a study of very simple nonlinear difference equations ... should be a part of high school or elementary college mathematics courses. They would enrich the intuition of students who are currently nurtured on a diet of almost exclusively linear problems.

May and Oster 1976

29.1   Introduction

The models of the previous parts of this book may be distinguished between those that are based on linear relationships among system components, such as the direct proportionality of the birth of fish in a lake modeled in Chap. 1, and those that capture nonlinearities, such as the per capita grass consumption by sheep modeled in Chap. 25. The models based on linear relationships could have been solved using analytical methods. In contrast, some of the nonlinear relationships discussed in this book have no analytical solutions. To solve for their behavior over time requires numerical solutions, such as the ones that we adopted here.

At least one feature, however, is common to both the linear and nonlinear models discussed so far. All of them have been "well behaved" in the sense that, at least in principle, we could have been able to predict their future time paths with relatively high confidence just by observing their past behavior and their current states. The models presented in this part of the book are different from these "well-behaved" models. Events occur that are not anticipated. Chaos prevails.

A save-disabled version of STELLA® and the computer models of this book are available at www.iseesystems.com/modelingeconomicsystems.

M. Ruth and B. Hannon, *Modeling Dynamic Economic Systems*,
Modeling Dynamic Systems, DOI 10.1007/978-1-4614-2209-9_29,
© Springer Science+Business Media, LLC 2012

29.2 Modeling Preference Cycles

The first of our chaos models returns to the case of the consumer who attempts to maximize his or her utility from the consumption of goods and services. That is the topic of this chapter. The following chapter concentrates on time lags in adjustments of demand and supply. Chapter 31 concentrates on price expectations and production lags, and Chap. 32 presents an example of chaos in macroeconomic models. Each of these models involves only a small number of state variables and controls. Yet, their dynamics are surprisingly rich.

We have modeled utility-maximizing consumers in the setting of a barter economy in Chap. 15. The model in Chap. 15 did not explicitly take into account the time over which the exchange of goods took place. Rather, the goods were exchanged among the participants in the trade, and only after an equilibrium in the trade occurred could consumption of the goods that were exchanged take place. In contrast, let us now explicitly model the history of consumption as a determinant of the utility that can be achieved by a consumer. Let us assume that the consumer's preferences for consumption goods are a function of the quantities of those goods consumed in the past. A discussion of endogenous preferences can be found in Rosser (1991).

We assume that the following utility function represents those preferences:

$$U = Q1^A {}^* Q2^{(1-A)}. \tag{29.1}$$

We also assume that the consumer wishes to maximize his or her utility function subject to the budget constraint

$$M = P1 {}^* Q1 + P2 {}^* Q2, \tag{29.2}$$

where $Q1$ and $Q2$ refer to the quantities of two goods, and $P1$ and $P2$ are their respective unit prices; M is the total budget available to purchase goods 1 and 2.

Before we introduce changes in consumption in response to past consumption, let us first find the optimum consumption choice in the static case. The static optimization of (29.1) subject to the constraint in (29.2) is equivalent to maximizing the following auxiliary function, frequently referred to as a *Lagrangian function*:

$$L = U + \lambda^* (M - P1 {}^* Q1 - P2 {}^* Q2), \tag{29.3}$$

where λ is a shadow price—the marginal utility of income. Maximization of (29.3) yields the following necessary conditions

$$\frac{\partial L}{\partial Q1} = A \frac{U}{Q1} - \lambda P1 = 0, \tag{29.4}$$

$$\frac{\partial L}{\partial Q2} = (1 - A) \frac{U}{Q2} - \lambda P2 = 0. \tag{29.5}$$

These conditions are also sufficient, in the case modeled here. They can be combined to

$$Q1 = \frac{A}{1-A}\frac{P2}{P1}Q2. \tag{29.6}$$

Compare this result to the optimal relationship between inputs into the production process of a profit-maximizing firm, derived in Chap. 8.

To calculate the demand curve for good 1, we replace $Q2$ in (29.6) with the use of the budget constraint of (29.2). The result is shown in the following equation:

$$Q1 = A\frac{M}{P1}. \tag{29.7}$$

Now that we derived the utility-maximizing demand curve for good 1, let us introduce the dynamic aspects of this problem. Assume

$$A(t+1) = \text{ALPHA}\,{}^*Q1(t)\,{}^*Q2(t) \tag{29.8}$$

and insert (29.8) into (29.7). ALPHA is the parameter that reflects the relevance of the consumption history for the current choice of $Q1$ and $Q2$. The larger ALPHA, the more prominent the choice of $Q1(t)$ and $Q2(t)$ is for the choice of $Q1(t+1)$ and $Q2(t+1)$. By cutting the direct connection of A from its own previous value, we are setting this system up for chaos. Rewriting (29.7) as

$$Q1(t+1) = A(t+1)\frac{M}{P1} \tag{29.9}$$

and inserting (29.8), we get

$$Q1(t+1) = \text{ALPHA}\,{}^*Q1(t)\,{}^*Q2(t)\frac{M}{P1} \tag{29.10}$$

which, when we make use of the budget constraint again, leads to

$$Q1(t+1) = \text{ALPHA}\,{}^*\frac{M}{P1}\left(\frac{M}{P2} - \frac{P1}{P2}{}^*Q1(t)\right). \tag{29.11}$$

Let us now introduce a simplification of our model by assuming $P1 = P2 = 1$. As a result

$$Q1(t+1) = \text{ALPHA}\,{}^*Q1(t)\,{}^*M\,{}^*[M - Q1(t)]. \tag{29.12}$$

Setting the prices of the two goods equal makes the model simpler without sacrificing the message that we wish to convey. You can easily relax this assumption later on.

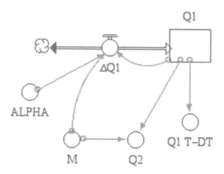

Fig. 29.1 Preference cycle model

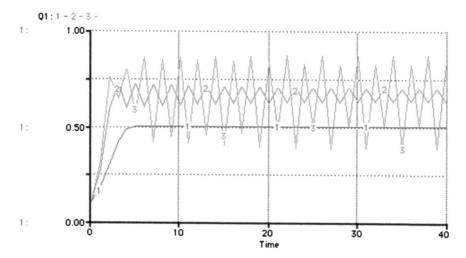

Fig. 29.2 Preference cycles with steady state and oscillations

Now that we defined demand in time period t, we can easily calculate the change in demand over time as

$$\Delta Q1 = Q1(t+1) - Q1(t) = \text{ALPHA} * Q1(t) * M * [M - Q1(t)] - Q1, \quad (29.13)$$

which we use to drive our model (Fig. 29.1).

The results illustrated in Fig. 29.2 are all calculated for $M = 1$. Let us vary the parameter ALPHA for consecutive runs, setting ALPHA $= 2$, ALPHA $= 3$, and ALPHA $= 3.5$. The change in demand over time is shown for the three cases in the following graph. Low values for ALPHA lead to a steady-state demand. Increases in ALPHA first lead to oscillations of period 1, then to oscillations of period 2.

The budget constraint is shown in Fig. 29.3. It is, as it should be, a straight-line curve, with a negative slope of $45°$, since we set $P1 = P2$.

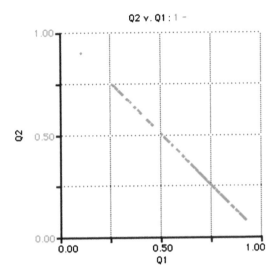

Fig. 29.3 Budget constraint

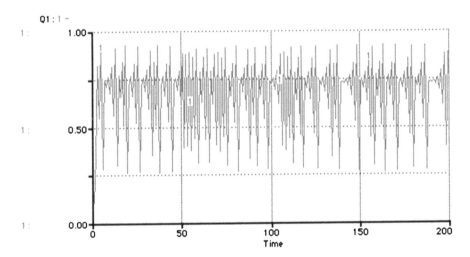

Fig. 29.4 Preference cycle with chaos

Let us increase ALPHA further to 3.7. The result is a breakdown of the regular periodic behavior of demand for good 1 (and hence for good 2). The model exhibits chaotic behavior (Fig. 29.4).

This behavior is exactly repeated from one model run to the next. It is not random, but deterministic. However, at each point in time we cannot predict the

Fig. 29.5 Phase diagram
for $Q1$

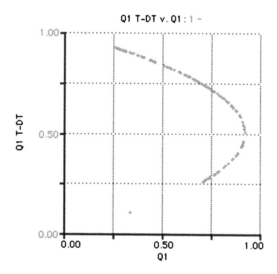

Fig. 29.6 Generating
randomness

future values of demand. All we do know is the range within which we will find the
state variable. Plot $Q1(t-1)$ against $Q(t)$—a so-called phase diagram—and you
will find that all the values of $Q1$ follow a well-defined pattern (Fig. 29.5).

Compare this result to a truly random number. We defined such a number in the
model as RAND and calculated its delayed value (Fig. 29.6).

The random number, plotted against its delayed value, is shown in the following
graph. Rerun the model several times and see how the graph changes. The difference between chaotic—but deterministic—behavior and random behavior should
become apparent (Fig. 29.7).

Set the DT of your model to 0.25 and vary the values of ALPHA to find the onset
of chaos. Then set DT $= 0.125$ and rerun the model. Interpret your result.

Replace (29.12) with the following demand curve

$$Q1(t+1) = M^*Q1^* \exp(\text{ALPHA}^*[1 - Q1(t)] \tag{29.14}$$

and find values for ALPHA that yield oscillations of period 1 and 2 and ultimately chaos. This equation was also used in a different context in May and
Oster (1976).

Fig. 29.7 Randomness

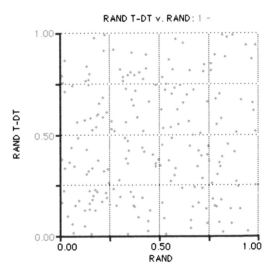

29.3 Preference Cycle Model Equations

Q1(t) = Q1(t - dt) + (ΔQ1) * dt
INIT Q1 = .1 {Units of Consumption Good 1}
INFLOWS:
ΔQ1 = (ALPHA*M^2 - ALPHA*M*Q1 - 1)*Q1 {Units of Consumption Good 1 per Time Period}
ALPHA = 2
M = 1 {Dollars}
Q1_T–DT = DELAY(Q1,DT,1)
Q2 = M-Q1 {Units of Consumption Good 2}
RAND = RANDOM(0,1)
RAND_T–DT = DELAY(RAND,DT)

References

May R, Oster G (1976) Bifurcations and dynamic complexity in simple ecological models. Am Nat 110:573–599

Rosser JB (1991) From catastrophe to chaos: a general theory of economic discontinuities. Kluwer Academic, Dordrecht

Chapter 30
Nonmonotonic Demand and Supply Curves

> *To these elementary laws [of nature] there leads no logical*
> *path, but only intuition, supported by being sympathetically*
> *in touch with experience.*
>
> Albert Einstein

30.1 Introduction

In Chap. 29, we saw chaos emerge in a model of a consumer's demand for a goods. In this chapter, we model chaos as a result of particular interactions between demand and supply of a good. Before we do this, let us briefly review the workings of a market characterized by demand and supply curves. Let us assume that demand and supply are not instantaneously equated with each other. Rather, adjustments in demand and supply take time.

Demand curves are typically assumed to be downward sloping because consumers are willing to pay less for a good that is available in abundance. The roots for the declining willingness to pay lie in declining marginal utility: Consumption of each additional unit of the good contributes less to the consumer's welfare the more of the good has already been consumed. Obvious examples include the consumption of water by someone walking in the desert. The first sip significantly contributes to that person's utility. The second sip will do so, too, but to a slightly smaller extent. After the first gallon, or so, each additional sip makes only a minor, and rapidly declining, contribution to utility.

Supply curves are typically assumed to be upward sloping. The underlying rationale is that producers experience decreasing returns to scale that lead to ever higher unit production costs as production is expanded.

A save-disabled version of STELLA® and the computer models of this book are available at www.iseesystems.com/modelingeconomicsystems.

M. Ruth and B. Hannon, *Modeling Dynamic Economic Systems*,
Modeling Dynamic Systems, DOI 10.1007/978-1-4614-2209-9_30,
© Springer Science+Business Media, LLC 2012

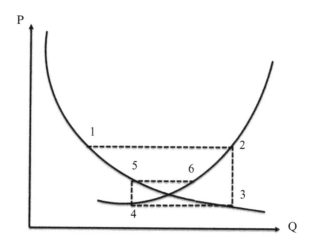

Fig. 30.1 Cobweb model

Under the assumptions of "normal" demand and supply curves, and perfectly functioning markets, any excess supply should lead to a reduction in demand, followed by a response in supply, until market equilibrium is achieved. This is illustrated in Fig. 30.1. The price of the good is denoted by P and quantities are noted as Q. The demand and supply curves are D and S, respectively.

For example, the model may start with an excess demand in point 1. As a result of excess demand, price is high, stimulating supply to rise to point 2, which leads to a subsequent drop in price, followed by movement to point 3, then 4, and so on, until demand and supply are equal. This model of the market clearing process is frequently referred to as a *cobweb model*.

Demand and supply curves, however, need not be as well behaved as we assumed previously. There are markets, such as for some highly skilled labor and in some fisheries that are assumed to have backward-bending supply curves. The origin for this "abnormality" can come from lags in the adjustment process among suppliers. For example, it takes time for workers to increase their skills before they can participate in the market for specialized workers. Similarly, it takes time to build boats and hire and train a crew of fishermen before one can go out to sea and successfully fish. In each of those cases, increasing prices may lead temporarily to a decline in supply—workers decide to withdraw from the labor market and participate in training programs; fishermen decide to build boats or nets and increase their knowledge of the sea.

30.2 Nonmonotonic Supply

An example for a system with "normal" demand curve and nonmonotonic supply curve is shown in Fig. 30.2. The result may be a chaotic cobweb, in which price does not settle down over time but keeps fluctuating in a deterministic, chaotic way.

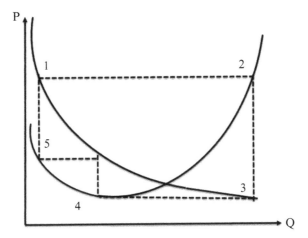

Fig. 30.2 Cobweb with nonmonotonic supply curve

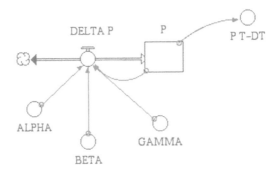

Fig. 30.3 Price dynamics in the cobweb with nonmonotonic supply

Due to the sensitivity of the price path to initial conditions, the future path cannot be predicted, although it will be found in a well-defined range.

A set of demand and supply curves that illustrate the chaotic cobweb model is the following (Rosser 1991):

$$Q(t) = S[P(t-1)] = A + B{}^{*}P(t-1) - E{}^{*}P(t-1)^{\wedge}2, \qquad (30.1)$$

$$Q(t) = D[P(t)] = C - D * P(t). \qquad (30.2)$$

To ensure that the market clears, the supply and demand quantities must be equal. The market-clearing equation for this system of demand and supply equations is

$$P(t) = \text{ALPHA} - \text{BETA}{}^{*}P(t-1) + \text{GAMMA}{}^{*}P(t-1)^{\wedge}2, \qquad (30.3)$$

where $\text{ALPHA} = (C-A)/D$, $\text{BETA} = B/D$, and $\text{GAMMA} = E/D$. The corresponding STELLA model is shown in Fig. 30.3.

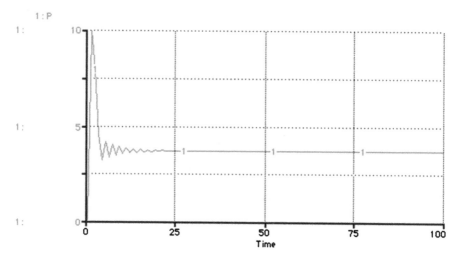

Fig. 30.4   Dynamic cobweb with BETA = 2.6

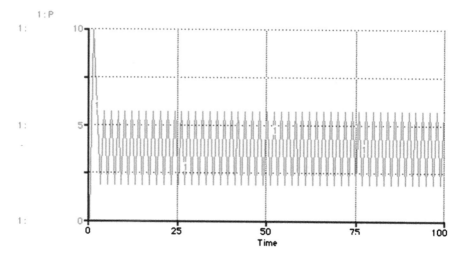

Fig. 30.5   Dynamic cobweb with BETA = 2.8

For the graph in Fig. 30.4 we set ALPHA = 10, GAMMA = 0.24, and BETA = 2.6. The result is a damped oscillation. In this case, the slope of the supply curve is such that prices gravitate toward the equilibrium, leading to market clearing. The graph in Fig. 30.5 is for BETA = 2.8. Here, we have oscillation with period 1. With each iteration, we neither get closer to the equilibrium price, nor do we move further away from it. In contrast to this result, BETA = 3.0 alternately leads to movement away from the equilibrium price and then again closer to it (Fig. 30.6).

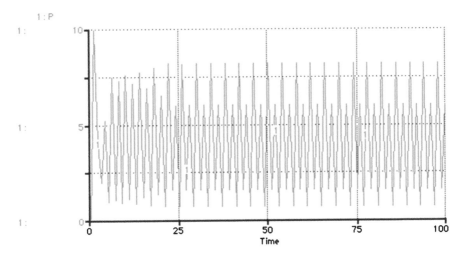

Fig. 30.6 Dynamic cobweb with BETA = 3.0

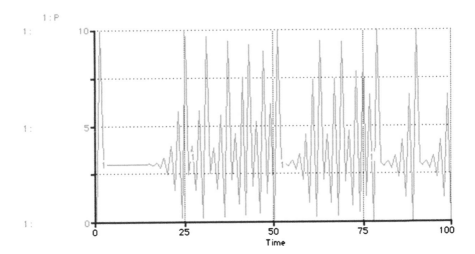

Fig. 30.7 Dynamic cobweb with BETA = 3.1

The last graphs in Figs. 30.7 and 30.8 are for BETA = 3.1 and exhibit chaotic behavior.

Set up a model with "normal" demand and supply curves to see the cobweb process in action. Then introduce a nonmonotonic demand curve. What could be reasons for the existence of such a demand curve and in what kinds of markets would you expect to find it? Can you generate chaos with your demand curve?

Fig. 30.8 Phase diagram for
BETA = 3.1

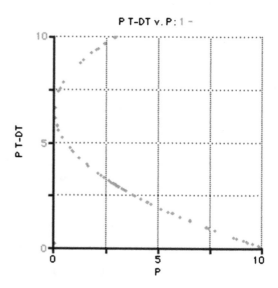

30.3   Nonmonotonic Supply Model Equations

P(t) = P(t - dt) + (DELTA_P) * dt
INIT P = .01 {Dollars}
INFLOWS:
DELTA_P = ALPHA-(1+BETA)*P+GAMMA*P^2 {Dollars per
Time Period}
ALPHA = 10
BETA = 2.6
GAMMA = .24
P_T–DT = DELAY(P,DT,.1)

Reference

Rosser JB (1991) From catastrophe to chaos: a general theory of economic discontinuities. Kluwer
    Academic Publishers, Dordrecht, The Netherlands, pp 46–47

Chapter 31
Price Expectation and Production Lags

All the effects of nature are only the mathematical
consequences of a small number of immutable laws

P. Laplace, in E. Bell, *Men of Mathematics*,
1937, p. 172.

31.1 Chaos with Price Expectations and Production Lags

In the Chap. 30, we saw how nonmonotonic demand and supply curves can cause
chaotic price behavior in dynamic cobweb models. In this chapter, we introduce
price expectations and lagged production into a model of *monotonic* demand and
supply curves. This model follows ideas laid out in Chiarella (1988).

The demand curve is a linear function of observed prices $P(t)$

$$Q(t) = D[P(t)] = A - B^*P(t). \tag{31.1}$$

The supply curve, in contrast, is a nonlinear function in expected prices.
The quantity $Q(t)$ supplied in period t is determined in the following way. In the
previous period $t-1$, firms formed expectations about the price of the good in period t.
That expected price is $P'(t)$. Based on this price, firms produce the quantity $Q(t)$ that is
offered on the market. Thus, we have here a combination of production lags and price
expectations that give rise to the following general supply curve:

$$Q(t) = S[P'(t)]. \tag{31.2}$$

A save-disabled version of STELLA® and the computer models of this book are available at
www.iseesystems.com/modelingeconomicsystems.

M. Ruth and B. Hannon, *Modeling Dynamic Economic Systems*,
Modeling Dynamic Systems, DOI 10.1007/978-1-4614-2209-9_31,
© Springer Science+Business Media, LLC 2012

Fig. 31.1 Supply curve

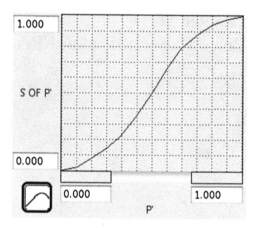

For our model, this supply curve is assumed to have an inflection point. The supply curve is specified by the graph in Fig. 31.1:

Assume that price expectations $P'(t + 1)$ are formed according to the following rule

$$P'(t+1) = (1 - W)^* P'(t) + W^* P(t). \tag{31.3}$$

The parameter W relates the price that was expected for the current period t with the price that is actually observed on the market in that period to form a new price expectation. A linear combination of both results in the newly expected price for the following period. For example, if $W = 0.5$ and the firms expect a price $P'(t) = 2$ but observe a price $P(t) = 1$, then they adjust their expectations for the following period to $P'(t + 1) = 1.5$. Thus, the parameter W captures the speed of the adjustment in price expectations.

Let us express $P'(t + 1)$ as a function of $P'(t)$. To do this, we solve (31.1) for $P(t)$ to get

$$P(t) = \frac{Q(t) - A}{B}. \tag{31.4}$$

Then, we insert equation (31.4) into (31.3) which yields

$$P(t + 1) = (1 - W)^* P'(t) + W^* \frac{Q(t) - A}{B}. \tag{31.5}$$

From this, we replace $Q(t)$ by using (31.2). We do this under the assumption that markets must clear. The result are expectations of future prices based on the expectations that were held for the prices of the current period and the supply that resulted from the price expectations for the current period:

$$P(t + 1) = (1 - W)^* P'(t) - \frac{A^* W}{B} + W^* \frac{S(P(t))}{B}. \tag{31.6}$$

Fig. 31.2 Price expectations and production lags

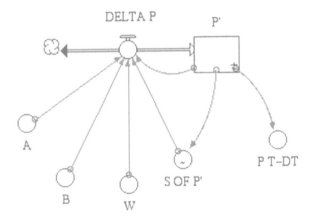

DELTA P

P'

A

B   W

S OF P'

P T–DT

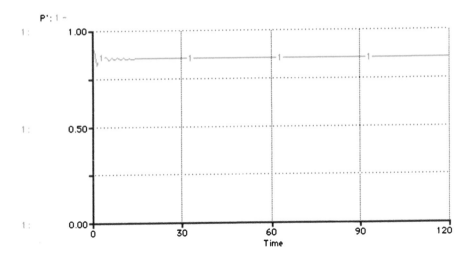

P': 1 –

Fig. 31.3 Price expectations and productions lags with $W = 3.2$

The resulting model for (31.6) is shown in the STELLA diagram of Fig. 31.2.

We set $A = 0.1$, $B = 1.0$, and vary W to find chaos. The choice of $W = 3.2$ results in damped oscillations. Increasing W, for example, to $W = 5$, $W = 5.5$, and $W = 5.75$, we get oscillations with one, two, and higher periods (Figs. 31.3–31.6). Increasing W further ultimately leads to the breakdown of the regular periodicity and the emergence of chaos.

The last set of graphs (Figs. 31.7 and 31.8) have been calculated for $W = 6.2$. Here, we have chaos. Run the model for a few periods, then pause it and make a guess on the future path of $P'(t)$. Any pattern that seems to emerge for the time path is soon abandoned. Yet, the path is not random. There is an underlying order, as is apparent

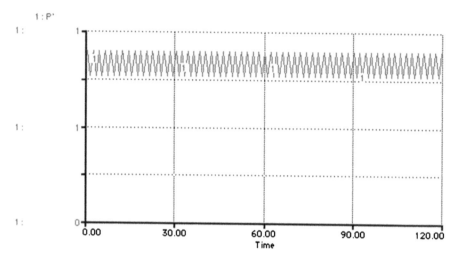

Fig. 31.4 Price expectations and productions lags with $W = 5.0$

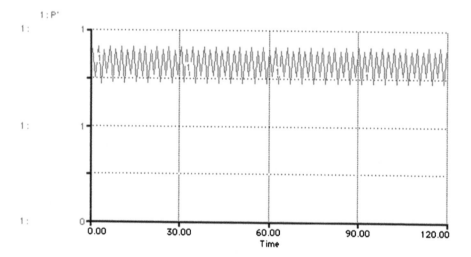

Fig. 31.5 Price expectations and productions lags with $W = 5.5$

from the second graph that follows. Can you find chaos by setting $0 < W < 1$ and varying the other parameters?

In the Chap. 30, chaos was generated by making use of a U-shaped supply curve. In this chapter, the use of a supply curve that is convex for low prices and concave for high prices enables us to generate chaotic behavior of the state variable. In the next chapter, we model production and investment from a macroeconomic perspective and, again, find chaos in a rather simple model.

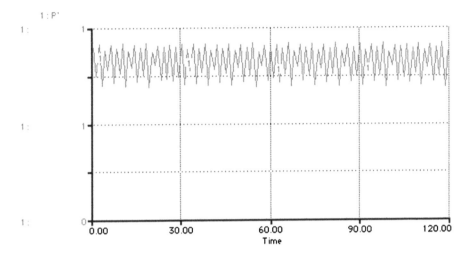

Fig. 31.6 Price expectations and productions lags with $W = 5.75$

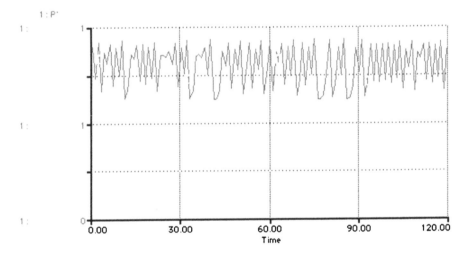

Fig. 31.7 Price expectations and productions lags with $W = 6.2$

Fig. 31.8 Phase diagram for
$W = 6.2$

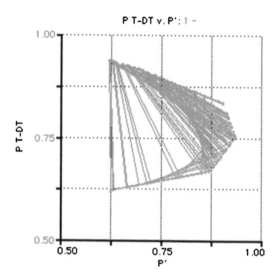

31.2 Price Expectations and Production Lag Model Equations

P'(t) = P'(t - dt) + (DELTA_P) * dt
INIT P' = .9 {Dollars}
INFLOWS:
DELTA_P = -A*W/B - W*P' +W*S_OF_P'/B   {Dollars per Time
Period}
A = .1
B = 1
P_T–DT = DELAY(P',DT,.7)   {Dollars}
S_OF_P' = GRAPH(P')
(0.00, 0.00), (0.0833, 0.02), (0.167, 0.08), (0.25, 0.145),
(0.333, 0.23), (0.417, 0.355), (0.5, 0.5), (0.583, 0.66), (0.667,
0.8), (0.75, 0.88), (0.833, 0.945), (0.917, 0.98), (1.00, 1.00)
W = 3.2

Reference

Chiarella C (1988) The cobweb model: its instability and onset of chaos. Economic modelling
    5:377–384

Chapter 32
Chaos in Macroeconomic Models

> *Earliest men perceived ... their world as largely chaotic, and so now do we! There is a difference, however. The cavemen viewed nature as indifferently rolling unbiased dice; modern men recognize that nature's dice are only slightly but nonetheless purposefully loaded.*
>
> Joseph Ford

32.1 Macroeconomic Chaos

Assume that the output of the economy is given by

$$F = B \, {}^* K^\wedge \text{BETA} \, {}^* (M - K)^\wedge \text{MU}, \tag{32.1}$$

where B, BETA, M, and MU are positive numbers; F is output per capita; and K is the capital–labor ratio of the economy. The difference $M-K$ represents the negative effects of capital concentration, such as pollution.

Allow the capital–labor ratio to change over time as

$$K(t + 1) = (F - H)/(1 + \Omega), \tag{32.2}$$

where H is the per capita consumption and Ω is the population growth rate. Furthermore, define

$$H = (1 - S) \, {}^* F. \tag{32.3}$$

A save-disabled version of STELLA® and the computer models of this book are available at www.iseesystems.com/modelingeconomicsystems.

M. Ruth and B. Hannon, *Modeling Dynamic Economic Systems*,
Modeling Dynamic Systems, DOI 10.1007/978-1-4614-2209-9_32,
© Springer Science+Business Media, LLC 2012

Fig. 32.1 Macroeconomic
chaos

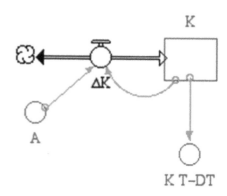

When all these equations are combined, we have

$$K(t+1) = A * K^{\wedge}\text{BETA} * (M - M)^{\wedge}\text{MU}, \tag{32.4}$$

where

$$A = S * B / (1 + \Omega). \tag{32.5}$$

For STELLA we need to calculate the change in K over time. This is given by

$$\Delta K = K(t+1) - K(t) = A * K^{\wedge}\text{BETA} * (M - K)^{\wedge}\text{MU} - K. \tag{32.6}$$

Let us simplify (32.6) by assuming BETA $= 1$, $M = 1$, and MU $= 1$. We relax this assumption later. The STELLA model is shown in Fig. 32.1. Related macroeconomic chaos models and an extensive discussion can be found in Jenson (1987).

Run the model with an initial K of 0.1, DT $= 1$, and $A = 1$. Then, for subsequent runs, raise A to 2 and 3 and watch K. You will find that for $A < 3.0$ there is one solution, namely, the logistic. The graph in Fig. 32.2 confirms this.

For $3.0 < A < 3.4$ there are two solutions; for $3.4 < A < 3.5$ there are four solutions; for $A > 3.57$ there is essentially an infinite number of solutions (chaos) with notable exceptions (try $A = 3.83$). The upshot of this model is that, as A rises to 3.57 (and DT $= 1$), increasing bifurcation is seen but the number of solutions, for K is finite. For $A \geq 3.57$, one cannot predict what the final equilibrium value will be or even whether there will be a finite or infinite number of solutions: There is no definite set of equilibrium values. However, some values seem to be "visited" more frequently than others. The graphs in Figs. 32.3 and 32.4 are plotted for $A = 3.7$ at DT $= 1$.

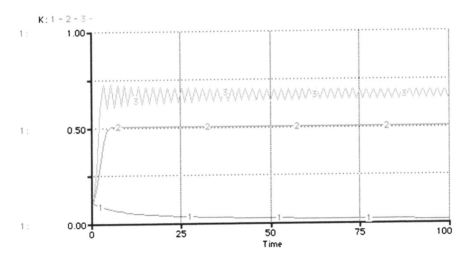

Fig. 32.2 Capital to labor ratio, K, for different values of A

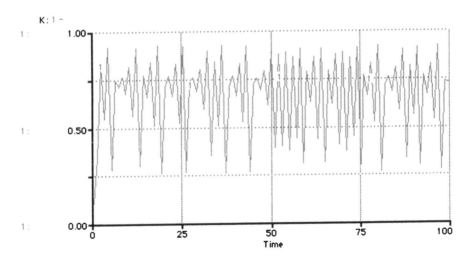

Fig. 32.3 Capital to labor ratio, K, for $A = 3.7$

Now, lower the time step to DT $= 0.5$ and find the value for A at which chaos begins again. Keep shortening DT, and you will find that there is a relation between the size of DT and the smallest A necessary to produce chaotic behavior:

DT	A necessary for chaos
1	3.58
0.5	6.12
0.25	11.29
0.125	21.56
0.0625	42.13
0.03125	83.24

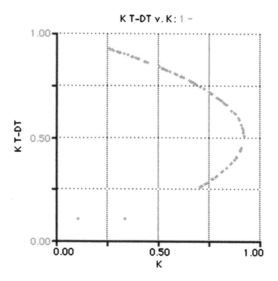

Fig. 32.4 Phase diagram of capital to labor ratio, K, for $A = 3.7$

A pattern emerges. When DT is halved, A is doubled and lessened by 1. That is, if $DT(n + 1) = 0.5*DT(n)$, then $A(n + 1) = 2*A(n)-1$. A function $A(DT)$ to calculate the critical A is

$$A(DT) = (4.57/DT) - (1 + 2 + 4 + 8 + ...1/DT) \qquad (32.7)$$

or

$$A(DT) = [A(1) + 1)/DT] - (1 + 2 + 4 + 8 + ...1/DT), \qquad (32.8)$$

the boundary between chaos and finitely numbered solutions for the spectrum of discrete steps.

This model shows you that the A for $DT = 0$ is infinity. This result is correct since chaos does not exist at the continuous level in a first-order differential equation. Chaos occurs on a continuous level only if you are stuck with a specific DT in your particular problem and the parameters lie within the critical range.

Now let us return to (32.6) without making the simplifying assumptions that gave rise to the previous results. Assume, for example,

$$\Delta K = [(A*(K^{\wedge}.7)*(1 - K)^{\wedge}.8) - K]/DT \qquad (32.9)$$

and find values for A that, alternatively, yield damped oscillations, cycles with one or two periods and chaos. Run the model first at a $DT = 1$, then cut DT in half and find the critical value for A at which chaos emerges. Can you find a relationship between the DT and the critical A?

32.2   Macroeconomics Chaos Model Equations

```
K(t) = K(t - dt) + (ΔK) * dt
INIT K = .1
INFLOWS:
ΔK = ((A*(K^1)*(1 - K)^1) - K)/dt
A = 3.7
K_T–DT = DELAY(K,DT)
```

Reference

Jenson RV (1987) Classical chaos. Amer Sci 75:168–181

Part VII
Conclusion

Chapter 33
Building a Modeling Community

The models and concepts that we developed in this book are powerful means to investigate the behavior of systems. The modeling approach that we chose is dynamic with regard to four issues. First, the systems that we modeled are dynamics ones, and we portrait their dynamics, rather than use a static or comparative-static approach. Second, the model development process itself is dynamic. We encourage you to start with simple models of complex economic systems. You will soon find that your models become increasingly complex but more representative of the system about which you are asking important questions. STELLA through its use of graphics is an excellent tool to organize and assess the various aspects of the system that you wish to capture. Once the model system is sufficiently understood, you can easily move on to expand on the model and capture additional features of a real economy.

Third, the learning process that accompanies model development and model runs is a dynamic one. By carefully phrasing the questions that the model should answer and by stating the assumptions that underlie the model that we develop, we structure our knowledge about a system. By running the model and observing the results, we learn about some aspect of a system. Subsequent model runs and model refinements provide more of this insight and should sharpen our focus for future model development and improve our intuition about the behavior of dynamic systems.

Fourth, through building and running computer models we provide a basis for communicating data and model assumptions. Frequently, model efforts become large-scale multidisciplinary endeavors. STELLA is sufficiently versatile to enable development of complex, large-scale dynamic models. Such models can include a variety of features that typically are not dealt with by an individual modeler. Through easy incorporation of new modules into existing dynamic models and flexibility in adjusting models to specific real-world problems, STELLA fosters dialogue and collaboration among modelers. It is a superb organizing and

---

A save-disabled version of STELLA® and the computer models of this book are available at www.iseesystems.com/modelingeconomicsystems.

M. Ruth and B. Hannon, *Modeling Dynamic Economic Systems*,
Modeling Dynamic Systems, DOI 10.1007/978-1-4614-2209-9_33,
© Springer Science+Business Media, LLC 2012

knowledge-capturing device for model building in an interdisciplinary arena. Individuals can easily integrate their knowledge into a STELLA model without "losing sight" of, or influence on, their particular part of the model. We anticipate that the modeling approach presented in this book will increase interaction among modelers and will be generating new momentum for interdisciplinary and cross-cultural exchange of ideas.

With the books in this series, we wish to initiate a dialogue with (and among) you and other modelers. We invite you to share with us your ideas, suggestions, and criticisms of the book, its models and its presentation format. We also encourage you to send us your best STELLA models. We intend to make the best models available to a larger audience, possibly in the form of new books, acknowledging you or your group as one of the selected contributors. The models will be chosen based on their simplicity and their application to an interesting phenomenon or real-world problem.

Index